Inheritance in Poultry

by Charles B. Davenport

with an introduction by Jackson Chambers

This work contains material that was originally published in 1906.

This publication is within the Public Domain.

This edition is reprinted for educational purposes and in accordance with all applicable Federal Laws.

Introduction Copyright 2018 by Jackson Chambers

Introduction

I am pleased to present yet another title on Poultry.

The work is in the Public Domain and is re-printed here in accordance with Federal Laws.

As with all reprinted books of this age that are intended to perfectly reproduce the original edition, considerable pains and effort had to be undertaken to correct fading and sometimes outright damage to existing proofs of this title. At times, this task is quite monumental, requiring an almost total "rebuilding" of some pages from digital proofs of multiple copies. Despite this, imperfections still sometimes exist in the final proof and may detract from the visual appearance of the text.

I hope you enjoy reading this book as much as I enjoyed making it available to readers again.

Jackson Chambers

TABLE OF CONTENTS.

	Page
A. Statement of Problem	1
B. Method and Material	5
C. Results of Crossing	6
Series I. Single-comb Black Minorca and White-crested Black Polish	6
Statement of Problem	6
The Races as a Whole	6
Table of Characteristics	6
Remarks on the Characteristics	7
1. Comb, 7; 2. Nostrils, 7; 3. Form of Skull, 8; 4. Crest, 9; 5. Color of Crest, 10.	
Material	10
Results	10
1. Comb, 10; 2. Nostrils, 12; 3. Cerebral Hernia, 13; 4. Crest, 14; 5. Color of Top of Head, 15; 6. Correlation of Characters, 16.	
Conclusions	17
Series II. Single-comb White Leghorn and Houdan	18
Statement of Problem	18
The Races as a Whole	18
Table of Characteristics	19
Discussion of Characteristics	19
1. General Plumage Color, 19; 2. Color of Upper Mandible, 19; 3. Nostrils, 19; 4. Comb, 19; 5. Whiskers or Muff, 20; 6. Beard, 20; 7, 8. Crest; Cerebral Hernia, 20; 9. Foot Color, 20; 10. Number of Toes, 20.	
Previous Investigations	21
Material	21
Results	21
1. General Plumage Color, 21; 2. Color of Upper Mandible, 22; 3. Nostrils, 22; 4. Comb, 22; 5. Face Feathering, 23; 6. Beard, 24; 7. Cerebral Hernia, 24; 8. Crest, 24; 9. Foot Color, 24; 10. Number of Toes, 25; 11. Correlation of Characters, 25.	
Conclusions	26
Series III. Houdan and Single-comb Black Minorca	27
Statement of Problem	27
The Races as a Whole	27
Table of Characteristics	27
Material	28
Results	28
1. General Plumage Color, 28; 2. Comb, 28; 3. Nostril Form, 28; 4. Crest, 28; 5. Cerebral Hernia, 28; 6. 7. Muff and Beard, 28; 8. Foot Color, 28; 9. Toes, 28.	
Conclusions	28

	Page
Series IV. Single-comb White Leghorn and Rose-comb Black Minorca	29
Statement of Problem	29
The Races as a Whole	29
Table of Characteristics	29
Remarks on the Characteristics	29
1. General Plumage Color, 29; 2. Comb Form, 29; 3. Foot Color, 29.	
Material	29
Results	30
1. Plumage Color, 30; 2. Comb Form, 30; 3. Foot Color, 31.	
Conclusions	31
Series V. Single-comb Black Minorca and Dark Brahma	31
Statement of Problem	31
The Races as a Whole	31
Table of Characteristics	32
Remarks on the Characteristics	32
1. General Plumage Color, 32; 2. Wing Bars, 32; 3. Comb, 32; 4. Earlobe Color, 33; 5. Iris Color, 33; 6. Foot Color, 33; 7. Foot Feathering, 34; 8. Vulture Hock, 34.	
Material	34
Results	34
1. General Plumage Color, 34; 2. Wing Coverts, 35; 3. Comb, 35; 4. Earlobe Color, 35; 5. Iris Color, 35; 6. Beak and Foot Color, 35; 7. Foot Feathering, 35; 8. Vulture Hock, 35.	
Conclusions	36
Series VI. White Leghorn and Dark Brahma	36
Statement of Problem	36
The Races as a Whole	36
Table of Characteristics	36
Remarks on the Characteristics	36
1. Hackle Color, 36; 3. Wing Bow, 37.	
Material	37
Results	37
1. General Plumage Color, 37; 3. Wing Coloration, 37; 4. Tail Color, 38; 5. Comb Form, 38; 6. Earlobe, 38; 7. Iris Color, 38; 8. Vulture Hock, 38; 9. Foot Feathering, 38.	
Conclusions	38
Series VII. Black Cochin Bantam and White Leghorn Bantam	39
Statement of Problem	39
The Races as a Whole	39
Table of Characteristics	39

TABLE OF CONTENTS.

	Page
Remarks on the Characteristics	39
1. General Plumage Color, 39; 2. Earlobe Color, 39; 3. Vulture Hock, 39.	
Material	39
Results	40
1. General Plumage Color, 40; 2. Earlobe Color, 40; 3. Vulture Hock, 40; 4. Foot Feathering, 40.	
Conclusions	40
Series VIII. White Leghorn Bantam and Buff Cochin Bantam	40
Statement of Problem	40
The Races as a Whole	40
Table of Characteristics	41
Remarks on the Characteristics	41
1. General Plumage Color	41
Material	42
Results	42
1. General Plumage Color, 42; 2. Earlobe Color, 43; 3. Vulture Hock, 43; 4. Foot Feathering, 43.	
Conclusions	43
Series IX. Tosa Fowl (Yokohama) and White Cochin Bantam	43
Statement of Problem	43
The Races as a Whole	43
Table of Characteristics	44
Remarks on the Characteristics	44
1. General Plumage Color, 44; 2. Tail, 44; 3. Foot Feathering, 48; 4. Foot Color, 48.	
Material	48
Results	49
1. General Plumage Color, 49; 2. Tail Length, 49; 3. Foot Feathering, 50; 4. Foot Color, 50; 5. Correlation of Characteristics, 51.	
Conclusions	51
Series X. Dark Brahma and Tosa Fowl	51
Statement of Problem	51
The Races as a Whole	51
Table of Characteristics	52
Remarks on the Characteristics	52
1. Shafting, 52; 2. Hackle Lacing, 52; 3. Body Lacing, 52; 4. Penciling, 53; 5. Red Wing-Bar, 53; 6. White Wing-Bows, 53; 8. White Earlobe, 53; 9. Iris Color, 53.	
Material	53
Results	54
1. Shafting, 54; 2. Hackle Lacing, 54; 3. Body Lacing, 54; 4. Penciling, 54; 5. Red Wing-Bar, 54; 6. White Wing-Bow, 54; 7. Comb, 54; 8. Earlobe Color, 54; 9. Iris Color, 54; 10. Foot Color, 54; 11. Vulture Hock, 54; 12. Foot Feathering, 55; 13. Tail Feathers, 55.	

	Page
Conclusions	55
Method of Inheritance, 55; Sex in Inheritance, 55.	
Series XI. Frizzle and Silky	55
Statement of Problem	55
The Races as a Whole	55
Table of Characteristics	57
Remarks on the Characteristics	57
1. Plumage Color, 57; 2. Comb Form, 57; 3-5. Feather Form, 57; 6. Number of Toes, 59; 7. Skin Color, 59.	
Material	59
Results	59
1. Plumage Color, 59; 2. Comb, 59; 3-5. Curving of Shaft, Barb Length, and Barb Form, 59; 6. Number of Toes, 60; 7. Skin Color, 60; 8. Crest, 60.	
Conclusions	60
Series XII. Single-comb White Leghorn Bantam and Black-breasted Red Rumpless Game	61
Statement of Problem	61
The Races as a Whole	61
Table of Characteristics	61
Remarks on the Characteristics	61
4. Uropygium	61
Material	62
Results	62
1. General Plumage Color, 62; 2. Beak Color, 63; 3. Uropygium, 63; 4. Foot Color, 63.	
Conclusions	63
Series XIII. Black Cochin Bantam and Black-breasted Red Rumpless Game	63
Statement of Problem	63
The Races as a Whole	63
Table of Characteristics	63
Material	63
Results	64
1. General Plumage Color, 64; 2. Uropygium, 64; 3. Iris Color, 64; 4. Vulture Hock, 64; 5. Foot Feathering, 64.	
Conclusions	64
D. General Discussion	65
Inheritance of Particular Characteristics	65
Comb Form	65
Nostril Form	68
Cerebral Hernia	69
Crest	69
Whiskers, or Muff	70
Beard	70
Feather Form	70
Uropygium	71
Tail-Length	71
Vulture Hock	71
Foot Feathering	72
Extra Toes	72
Skin Color	73
Mandible Color	74
Foot Color	74

TABLE OF CONTENTS.

	Page
Iris Color	74
Earlobe Color	74
General Plumage Color	74
White vs. Dark	75
Dominance of White	75
Barring	75
Andalusian Coloration	76
White vs. Buff	76
Black vs. Red	76
Color of Top of Head	76
Color of Hackles—Hackle Lacing	77
Wing Color—Red Wing Coverts	77
Tail Color	77
Shafting	77
Body Lacing	78
Penciling	78
General Topics in Inheritance	78
Unit Characters	78
Alternative, Particulate (Mosaic), and Blending Inheritance	81
Inheritance of Specific vs. Varietal Characteristics	82
Inheritance of Positive vs. Negative Varietal Characteristics	83
Inheritance of Old vs. New Characteristics	84
Dominance and Recessiveness	84
Dependence of Dominance on the Races Crossed	86
Prepotency and Dominance	87
Hybrid Forms	88
Reversion	90
Purity of Gametes	91
Comparison of Reciprocal Crosses	93
Inheritance of Sexually Dimorphic Characteristics and Sexual Dimorphism in the Hybrids	93
Black Minorca and Dark Brahma	94
White Leghorn and Dark Brahma	94
White Leghorn and Houdan	94
White Leghorn and Rose-comb Black Minorca	95
Tosa Fowl and White Cochin Bantam	95
Dark Brahma (female) and Tosa Fowl (male)	95
Transfer of Sexually Dimorphic Characteristics from One Sex to the Other	95
Sex in Hybrids	97
Correlation of Characteristics	97
The Mutation Theory in its Relation to the Origin of Domesticated Animals	98
E. Summary of Conclusions	100
F. Literature Cited	101

INHERITANCE IN POULTRY.

By C. B. Davenport.

Evolution proceeds by steps. These steps are measured by the *characteristics* of organisms. When in the evolution of a race a characteristic is added a progressive step is taken. When a characteristic drops out a retrogressive step is made. Since the characteristic is the unit of evolution, it deserves careful study. The present work is a first study of the method of inheritance of characteristics.

A. STATEMENT OF PROBLEM.

When by some abnormal process a single, fertilized egg develops into two individuals they are, and continue throughout life to be, almost indistinguishable. This holds true even when the conditions of life of the two are dissimilar. This case is exemplified by "identical twins" as they occur in man.* The great similarity of such identical twins teaches that environment plays a small part in determining adult characteristics as compared with heredity. Consequently more confidence can be felt that the results of hybridization experiments are directly due to inheritance; they are little affected by varying environment.

The children of the ordinary family are not identical in appearance, although showing marked family traits. Certain characteristics may be common, but others are peculiar to each individual child. This proves that the fertilized eggs of the same two parents have not the same hereditary potentialities. It indicates also that we cannot predict the characteristics of the offspring from those of the parents. The proportion of qualities derived from either one of the two parents will differ in different children, or new qualities may appear. This is because the offspring do not inherit from the visible part of the parents' bodies but from their hidden germ cells or "gametes." And the characteristics of the soma are never throughout the same as those of the ripe gametes it carries.

When the parents belong to different races having markedly dissimilar characteristics there is not merely the question of dissimilarity of the offspring but of the inheritance of the antagonistic characteristics. Until

---

*Galton, F., 1883, pp. 216–243. Compare also for a critical study of resemblance in twins, Thorndike, 1905.

recently the law has been commonly accepted which is thus expressed by Darwin (1876, Chapter XV): "When two breeds are crossed their characters usually become intimately fused together." Many cases of non-fusing inheritance are now known and it is important to ascertain the relative frequency of the different kinds of inheritance and their relation to one another.

Lucas (1850, p. 194) recognizes three methods of inheritance, which he calls respectively that of election, of mixture, and of combination. They are thus defined: Election results in imprinting on some or all parts of the organism the characteristics of the father exclusively or those of the mother. Mixture results in a mixed or simultaneous representation of the father and of the mother on some or all of the parts of the organism. Its extreme is fusion of characteristics. Combination results in the substitution of a new characteristic in the place of any representative in a part or over the whole of the organism. This new characteristic results from the interaction of the two antagonistic ones just as a chemical combination often differs wholly from the elements which have been united in its manufacture.

Darwin (1876, Chapter XV) seems to recognize only two classes of inheritance, viz., one in which characteristics blend and one in which they refuse to blend. Of the latter class, however, there are two cases; either the hybrid receives all its characters from one of its parents only, or the hybrid receives part of its characters from one parent, the rest from the other.

Nägeli (1884; 1898, p. 17) describes the different forms of inheritance very clearly, thus:

> In the idioplasm of a germ cell arising from the crossing of unlike individuals the micellar rows of the individual Anlagen have sometimes an intermediate constitution and produce characteristics in the organism which are intermediate between the characteristics of the parents. Sometimes the micellar rows derived from the father and the mother respectively lie side by side interchanged in the idioplasm of the offspring in distinct groupings and may reproduce in the organism their respective characteristics side by side, or only one of them may develop, while the other remains latent. (Clark's translation.)

Galton (1889, pp. 7, 12, 14) distinguishes three kinds of inheritance, as follows: (1) *Particulate*, or inheritance "bit by bit, this element from one progenitor, that from another;" (2) *blending*, as in human skin color; this may "be none the less 'particulate' in its origin, but the result may be regarded as a fine mosaic too minute for its elements to be distinguished in a general view;" and (3) *exclusive*, as in human eye color; although "there are probably no heritages that perfectly blend or that absolutely exclude one another, but all heritages have a tendency in one or the other direction, and the tendency is often a very strong one."

The different types of inheritance are thought by various authors to be characteristic of particular sorts of crossing. Isidore Geoffroy St.-Hilaire insisted "that the transmission of characters without fusion occurs very rarely when species are crossed." De Vries (1905, pp. 253, 280) concludes that blending and particulate inheritance of qualities characterize the offspring of crossed species, whereas an alternative inheritance of qualities is characteristic of the offspring of a species crossed with a variety* or of two varieties crossed *inter se*.

In the case of alternative inheritance there often is exhibited an extremely suggestive phenomenon. When hybrids showing such inheritance are crossed *inter se* there is a segregation of the various alternative characteristics into different individual offspring. This is the discovery of Mendel (1866).† The attempt has naturally been made to generalize Mendel's law—to make it apply universally. In my own study the applicability of this law has been kept constantly in mind.

---

* It is to be recalled that in the De Vries system a variety differs from its parent species either in that a characteristic of the species has become latent in the variety or in that a characteristic which was latent in the species has reappeared in the variety. A new species, on the other hand, differs from its parent species in the acquisition of one or more wholly new characteristics. "In normal fertilization and in the intercrossing of varieties all characters are paired." Hence the paired characters struggle together in the zygote and the stronger one of the pair dominates or covers over the weaker one. Thus inheritance is alternative or exclusive. "In crosses between elementary species the differentiating marks are not mated." Hence there is no such struggle between characteristics; consequently those of both parents reappear in the offspring, interdigitating.

† The rediscovery of Mendel's work simultaneously by De Vries and by Correns in 1900 will always rank as one of the interesting coincidences in the history of science. There is evidence that others had independently discovered this law in their own work in the last third of the nineteenth century, but the history of this law is still to be written. I may note that Haacke, in 1893, as a result of extensive breeding of animals, expresses the law of purity of the germ cells. He has the theoretical idea that inheritance is conveyed both by the plasma (P) and the nucleus (Kern, K). In the union of dissimilar races two kinds of plasma (P and P′) and two kinds of nuclear material (K and K′) may be distinguished. On page 236 he says:

Die beiden verschiedenen Plasmen P und P′ die sich bei der Befruchtung vereinigt haben, trennen sich wieder bei der Reductionsteilung der Keimzelle, und dasselbe gilt von den beiden Kernstoffe K und K′. Diese Trennung ist in manchen Fällen, wie es scheint, eine völlige, so dass die Plasmen und die Kernstoffe, abgesehen von den mehr oder minder weitgehenden, aber niemals vollkommenen Ausgleichungen ihrer Eigenschaften, die durch gegenseitige Beeinflussung stattfinden müssen, ebenso rein aus der Vereinigung hervorgehen, als sie in diese hineingetreten sind.

Still further Haacke recognizes that in the separation of qualities that occur in the reduction period of the hybrid germ cells, those from different parents may gather into one germ cell. Since this occurs in accordance with the laws of chance (worked out in an example by Haacke), we have various combinations of characteristics in the second hybrid generation. Because of the purity of the germ cells it will often happen that mice having certain qualities will, when bred together, produce only those qualities, however complex their ancestry. For example, white dancing-mice bred together will produce nothing but white dancing-mice. Haacke's results seem to have been overlooked by recent experimenters.

In typical Mendelian cases not only do the qualities segregate in the second hybrid generation, but in addition, in the first generation, when two contrasted characteristics are bred together one of the two is patent in the offspring; the other does not appear. The first is the dominant quality; the second is recessive.

It seems at first to have been assumed that when one of two antagonistic characteristics was dominant over the other it was so in all cases. Recent studies have, however, greatly expanded our notion of dominance and recessiveness. Even in alternative inheritance we have to admit various additional phenomena of which the following are examples: *Prepotency* of a character, elsewhere recessive, in some individual or strain. *Latency*, as Castle (1905, p. 24) uses the term, or the inactive persistence of a normally dominant characteristic in a recessive individual or gamete. When the recessive is cross-bred the latent characteristic may appear as a dominant. *Reversion*, or the assumption of an atavistic character by a heterozygote. This is illustrated by the case of the cross between albino and black-and-white mice which throw gray. However, this instance may be one of latency. In this study attention will be paid to these phenomena.

What determines dominance in any case? This is a disputed point. De Vries (1905, pp. 278, 280) suggests "that hybrids between a species and its retrograde variety will bear the aspect of the species," and "that the older character dominates the younger one." However, he says it is not the systematic relation of the two parents of a cross that is decisive, but only the occurrence of the same quality, in the one in an active, and in the other in an inactive condition. Hence, whenever this relation occurs between the parents of a cross the active quality prevails in the hybrid, even when the parents differ from each other in other respects so as to be distinguished as systematic species. Correns (1905) also cites cases in which the active allelomorph dominates. In my studies constant attention is directed toward this matter.

To recapitulate: This study has been undertaken to determine the different forms of inheritance (alternative, particulate, blending) occurring in poultry, and to study especially the phenomena of alternative inheritance as exhibited in this group in order to see in how far they accord with Mendel's law and in how far the accessory phenomena of dominance, latency, and reversion occur.

B. METHOD AND MATERIAL.

To answer in the shortest time the foregoing questions about inheritance it was necessary to use some rapid and fecund breeder and to interbreed both varieties and species. But species-breeding is slower and more difficult and not more important than breeding races; for while on the one hand it may be urged that races are artificial, having arisen under domestication, on the other hand hybridization between established species probably plays little part in nature. What must occur again and again in nature is the mating of a mutation or newly arisen race with the parent species. It has been urged that, in such cases, the rare mutation must be swamped by intercrossing with the numerous representatives of the species. But if new characters do not blend in breeding we can see that a new characteristic once arisen may not be swamped. Consequently the study of inheritance in races assumes first importance, and domestic races afford the best material for such study.

Again, if we accept the doctrine that man is a single species, all the momentous questions of human inheritance are questions of race inheritance. The outcome of such an admixture of races as is going on in America is a question of race inheritance. The offspring of a man and a woman having one or more diverse characteristics will follow the laws deduced from a study of crossed races. These are practical problems of human evolution, and experiments made with domesticated races can throw light upon them.

The main material utilized has been, as stated, poultry. Poultry offer these great advantages: That they are easily bred in great numbers, that two generations can be reared to maturity in a year, that they stand much inbreeding without loss of fertility, and that the number of well-defined characteristics in the group is very great.

In my experiments I have kept 29 pens, each with its cock and one or more hens. To separate the eggs of the different females, trap nests were used in the later experiments to hold the hen until she is released. Her number is read and written, with date, upon the egg. Before placing the eggs in the incubator, a list is made of them. Before hatching, eggs of each of the different parentages are separated into a compartment by themselves, so that the exact parents of each chick may be known. A legband is applied to the chick the moment it is removed from the pedigree tray of the incubator. By these means I have gained in one year 1,500 offspring derived from known parents, and have reared about 500 of them to a period when their adult plumage characters were distinguishable. For keeping records I have used a field pocketbook and a day book at my work-table. "Loose leaf" forms were used for the description of each of the stock, for its egg record, for a chart of its plumage, and for its photographs. Finally, the results of each set of experiments are kept in a large book, posted nearly to date.

In treating of my experiments I propose first to give the results by the races crossed, and, secondly, to discuss in order the problems that were set at the beginning. My experiments led me to lay little stress on the races as named by fanciers. In fact it is not races that have been crossed but *characteristics*. However, as the breeders' names have a utility in bringing to the mind a certain combination of characteristics, they have been freely employed. The different races whose offspring are discussed in this paper are given below in the order adopted in Wright's Poultry Book.

Buff Cochin (Bantam).
Black Cochin (Bantam).
Dark Brahma (Bantam).
Black-breasted Red Game.
Single-comb Black Minorca.
Rose-comb Black Minorca.
Single-comb White Leghorn.

White-crested Black Polish.
Houdan.
Frizzle.
Silky.
Tosa fowl, or Yokohama.
Rumpless Game Bantam.

C. RESULTS OF CROSSING.

Series I.—Single-comb Black Minorca and White-crested Black Polish.

STATEMENT OF PROBLEM.

The cross was undertaken primarily to learn the method of inheritance of the crest, cerebral hernia, and comb of the remarkable Polish fowl.

THE RACES AS A WHOLE.

The Black-crested White Polish (figs. 1 and 2) belongs to a class—Polish fowl—which is one of the fundamental types of poultry. The origin of the Polish fowl is obscure. They were mentioned by Aldrovandi in the sixteenth century. They are found to-day in most parts of the world, and their most characteristic feature may, indeed, have originated independently many times. This feature is a cerebral hernia and its associated crest of large feathers.

The Single-comb Black Minorca (figs. 3 and 4) is a typical representative of the Mediterranean class of poultry—tall, stately, close-feathered, non-broody fowl,—modern representatives of the ancient Egyptian poultry. They seem, indeed, to have come from Spain, those imported to England having, according to Wright (1902, p. 391), come from the island whose name they bear. The single-combed form is the original and typical variety.

TABLE OF CHARACTERISTICS.

No.	Characteristic.	Single-comb Black Minorca.	White-crested Black Polish.
1	Comb	Single, very large	Two papillæ.
2	Nostrils	Narrow	Wide or high.
3	Top of head—skull	Normal	Cerebral hernia.
4	Top of head—plumage form	Plain	Crested.
5	Top of head—plumage color	Black	White and black.

REMARKS ON THE CHARACTERISTICS.

1. COMB.—The single comb of the Minorca (fig. 4) is derived from the primitive wild ancestor, for all the four feral species of Gallus have a single comb. What is remarkable about the comb of the Minorca is, however, its enormous size, gaining in many fowls a length of 150 mm. and a height of 100 mm. This seems to have been brought about by selection of extreme variants in fluctuating variability; at any rate, English breeders have gone further than American breeders have thought wise in the production of enormous combs.

The Polish comb (fig. 7, Pl. II) is a remarkable structure and is a phylogenetically new form. Some breeders try to eliminate it altogether; others retain it as a pair of horns. I quote from some of the "Standards" and descriptions of authors. Mr. P. Jones in Tegetmeier (1867, p. 176) says: "There should be no appearance of comb." Darwin (1876, Chapter VII) says: "Comb absent or small and of crescentic shape." According to Wright (1902, p. 443), "The comb should be practically absent, but on close inspection two very small horns can generally be discerned." Baldamus (1896, p. 149) states: "Kamm nicht oder kaum bemerklich, höchstens 2 kleine Spitzen zulässig." The American Standard of Perfection (1905, p. 152) announces: "Comb V-shaped, of small size, the smaller the better; set evenly on head, retreating into crest; natural absence of comb is preferred." My parental stock (fig. 7) had two large papillæ of irregular form and large confluent base. Further discussion of this type of comb will be deferred to page 65.

2. NOSTRILS.—In the Minorca, as in the Jungle fowl, each of the external nares is a horizontal slit bounded above and laterally by a fold of cornified skin (compare fig. 22). The slit leads into the outer ethmoid cavity. By removing the membrane the outer fold of the ethmoid can be seen as a ridge that extends well distad (Fig. A). In the Polish fowl the outer membrane

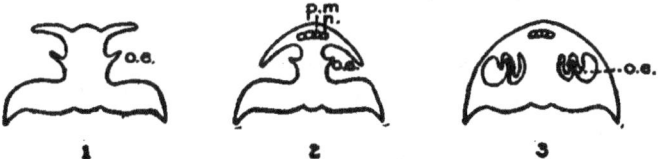

FIG. A.—1. Cross-section of beak of Polish fowl through wide nostril; o. e., outer ethmoidal fold. 2. Cross-section of beak of Minorca fowl through narrow nostril; p. m., premaxillary bone; n, nasal bone. 3. Cross-section of beak of fowl behind nostril; o. e., outer ethmoidal fold.

is so short that the narial aperture is very wide and the outer ethmoidal fold is exposed to view. The nostril may be said to be wide or "high" (figs. 7, 12). Sometimes the upper margin of the nostril may even be elevated above the level of the culmen of the beak, and in this case a transverse fold crosses the culmen from nostril to nostril. This I call the culminal fold.

The morphology of the nostril of the Polish fowl seems to be little understood. The term "cavernous" is applied to this form by the American

Standard of Perfection (1905, p. 13), which speaks of the hollow protruding nostrils. Wright (1902, p. 441) in treating of the cranial dome says: "Excess in one part being often connected with defect in some other, as Mr. Darwin pointed out, the skulls with this peculiarity usually show a chasm in the intermaxillary bones, which in other fowls support the roof of the nostrils; owing to which deficiency in bony support the nostrils of all heavily crested fowls appear flattened and depressed and yet cavernous in character." Darwin (1876, Chapter VII) attributes the width of the nostrils to the circumstance that the branches of the premaxillary and of the inner processes of the nasal bone are somewhat upturned.

Comparing the skull of a Polish fowl with that of a Minorca, I find the following relations:

First, the nasal bone has in the Minorca as in most other fowl the form of the Greek λ; the single stem (processus frontalis) projects caudad and lies as a flat plate above the frontal bone by which, also, it is cut off from contact with most of the lachrymal bone. The two anterior branches enclose between them the olfactory space. The processus maxillaris runs nearly perpendicular to the maxillary bone; the processus premaxillaris is a slender stylet terminating over the middle of the olfactory space and embracing the nasal process of the premaxillary bone (Fig. A, 2, n). In the Polish fowl the processus frontalis is relatively shorter and broader. The whole central nasal area is greatly depressed, forming a cavity in front of the cranial dome—a cavity that is filled with the cartilaginous foundation of the rudimentary comb. In front of this depression of the skull the processus premaxillaris rises, reaching about the normal height of this part of the nasal bone, and stops there in the posterior quarter of the nasal space. Thus the bony nasal space is posteriorly of normal height. What is peculiar in the skull of the Polish fowl is that the processes premaxillaris are very short and slender and do not embrace the nasal process of the premaxillary, but instead the cartilaginous dorsal edge of the orbitosphenoid or mesethmoid. This thickened dorsal edge continues anteriorly as the edge of the internasal septum, and it replaces the nasal processes of the premaxillary. Anteriorly the septum abuts upon the premaxillary. At this point there is adhering to the premaxillary a mass of tissue in the median plane which seems to be the rudiment of the nasal process of the premaxillary.

In criticism of Darwin's explanation of the wide nostril based on the ground that the nasal process is upturned, it may be said that the wide nostril lies in front of the upturned nasal bone and in a part of the nasal space that is not higher (nor wider) than in the normal skull. I think it must be concluded that the wide nostril is independent of the width of the nasal space. As we shall see later, the wide nostril is never found in connection with a single comb.

3. FORM OF SKULL.—Polish fowl have long been noted for the remarkable form of their skull. This was referred to by Bovelli (1670, cent. 2, p. 157,

teste Hagenbach) and has been studied by Blumenbach (1812), Hagenbach (1839), Tegetmeier (1856), Darwin (1876), and others. A dome rises from the front of the nasal bone often to a height of 15 or 18 cm. (figs. 7, 10, Pl. II). This dome is the secondary bony covering of a cerebral hernia whose dura mater has become ossified. Such cerebral hernias are not uncommon among poultry. Roughly, 1 per cent of the chicks (outside of the crested races) that failed to hatch in my incubations and were examined by me had such a hernia. In one or two instances Polish chicks that died before hatching were affected by incomplete closure of the cerebrum, the top of the head not being covered by bone or skin. Such an abnormality has been described by Hagenbach (1839, pp. 324–326) also. So profound a modification of the brain would naturally be associated with mental peculiarities. My own Polish have shown themselves very slow of movement, and two male Houdans (having the same sort of abnormal skull) were affected by some trouble in the head which led them to move backward, turn somersaults, and move otherwise abnormally. Hagenbach states that the Polish fowl are subject to apoplexy and epilepsy. It is remarkable that formerly the hens alone had the cerebral hernia (Darwin, 1876, Chap. VII), while now it occurs in both sexes.

4. CREST.—This consists of a number of large white feathers arising from the frontal region of the skull (figs. 1, 2, and 7). Structurally, they, like normal feathers from the top of the head, have the barbules of the distal portion of the barbs reduced so that the barbs do not interlock. This condition is seen in the hackles of all breeds. The form of the feathers resembles that of the hackle, being more attenuated in the male than the female. The great size of the crest feather, like that of the hackle, is largely due to its long period of growth. After molting, the new crest feathers are characterized by long and stout sheaths in which the feather develops. These persist after most of the other contour feathers no longer possess sheaths and consequently have stopped growing.

The cause of the crest is a matter of much interest. Hagenbach (1839, p. 329) raises the question and says:

Nicht ohne Bedeutung scheint mir übrigens eine auf die Hemicephalie bezügliche Beobachtung von Meckel* zu sein, welche so lautet: "Sehr merkwürdig ist die häufig vorkommende regelwidrige starke Entwicklung von Haaren an ungewohnten Stellen. So finde ich bei einigen von denen, welche ich vor mir habe, und gerade bei denen, wo der Hauptmangel am grössten ist, fast den ganzen Körper, besonders aber den Rücken, die Hüften und die obern Extremitäten, mit Haaren von 6 Linien bedeckt."

Whether or not the developmental disturbance is the cause of the prolonged growth of feathers, it is certain that the immediate cause is the unusual and prolonged nutrition of the feathers. The skin from which they arise is thick and rich in blood-vessels. Whether the cerebral hernia is a

---

*Handbuch der pathologischen Anatomie, Tom. I, p. 195–260.

necessary antecedent of the increased nutrition of the feathers or not can be tested by hybridization, which will show whether or not crest and dome are absolutely correlated.

5. COLOR OF CREST.—In our race this is white, in striking contrast with the rest of the plumage. This color is not necessarily associated with the crest. Wholly gold-spangled, silver-spangled, buff, white, and even black Polish have been created. On the other hand, since normal poultry with the top only of the head white are unknown, it seems probable that the color of the crest has dissociated itself from that of the other parts of the body as an independent unit character. Consequently it is not strange to find a black-crested white breed of Polish (Wright, 1902, p. 448). It will be of interest in our experiments to see how far color and crest are dissociable.

MATERIAL.

The mothers in this experiment were partly Polish, partly Minorca. The Polish mothers, Nos. 6 and 7,* were good representatives of their breed, with perfectly black plumage except for the well-developed white crest. The Minorca mothers, Nos. 13 and 14, were also typical birds without trace of mealiness in feathers.

The father Minorca (No. 12, fig. 4) had a great red comb, 150 mm. long, with seven points, one of which rose to 100 mm. above the level of the head. Its plumage was perfectly black. The paternal Polish (fig. 2) had as representative of the comb a pair of papillæ. The crest anteriorly contained some black feathers with white base or tip; next came feathers 75 to 100 mm. long, at first largely, then wholly white. At the posterior end of the crest, where it passes into the nape, black gradually makes its appearance until the exclusively black feathers of the neck are reached. White and black do not *blend* on any feather, but coexist in a *particulate* fashion.

RESULTS.

1. COMB.—*First generation.* Of 88 hybrids between single-combed and Polish-combed all follow a single type, which is, however, very variable. The comb is single anteriorly but bifurcated behind. This may be called the Y-shaped comb. The point of splitting occupies a variety of positions. Usually it lies in the middle third of the whole comb (fig. 8). In some cases, such as Nos. 67 ♂ (fig. 8), 176 ♀, 350 ♀, and 408 ♀, the splitting point is close to the anterior end, so that the comb is cup-shaped.† In other cases the point of division may be so far posteriad that only the last or the last few serrations are split. Indeed, in one case (No. 324, which died five weeks after hatching), the comb was apparently perfectly single. I regard this as the last term of the series and suspect that there existed, even in this

---

*Bought July 1, 1904; No. 7 died October 17, 1904.

†Such a comb is mentioned by Darwin (1876, Chap. VII) as formed when the two ends of a double comb are cemented together.

case, a repressed potentiality toward splitting of the comb at its posterior end. In general, then, the hybrid condition between single comb and supranasal papillæ is a Y-shaped comb* and there is an absence of dominance.

Second generation. When the birds with Y-shaped combs were interbred the 101 offspring were of three types, as follows:

Characteristic.	No. of individuals.	Percentages.		
		Actual.	Expected.	
			a	β
Single comb...........................	30	29.7	25	18.75
Cleft comb...........................	47	46.5	50	56.25
Papillæ or "absent"..................	24	23.8	25	25.00
Total...........................	101	100.0	100	100.00

Expectation in this case depends on the theoretical view we take of the nature of the unit characters involved. We may assume that single comb and V-comb are the allelomorphs and split comb a heterozygous type, constantly reproducing itself. On this assumption we should expect the proportion given in column a of the foregoing table. On the other hand, following the method of interpretation suggested by Cuénot (1903) and Correns (1905) in such cases, we may assume two pairs of allelomorphs, namely, (1a) median comb and (1b) no median, and (2a) no splitting and (2b) splitting. Taking median comb and splitting as dominant, the single comb combines the two characters: median comb, no splitting; the V comb combines: no median comb, splitting. In generation F_1 the four characters are combined: median (no median), splitting (no splitting), and the comb is Y-shaped, the characteristics put in parenthesis being recessive. The F_2 generation follows the law of inheritance in dihybrids:

		Per cent.	Class.	Relative frequency.
25 % (*Median* × *median*)..	Splitting × splitting .... 6¼ 2 (Splitting × no-splitting) .. 12½ } 18¾ % .. (a) No-splitting × no-splitting.. 6¼ .... (b)			3 1
50 % (*Median* × no-*median*)..................	Splitting × splitting .... 12½ 2 (Splitting × no-splitting) .. 25 } 37½ % .. (c) No-splitting × no-splitting.. 12½ .... (d)			6 2
25 % (No-*median* × no-*median*)..................	Splitting × splitting .... 6¼ 2 (Splitting × no-splitting) .. 12½ } 18¾ % .. (e) No-splitting × no-splitting... 6¼ .... (f)			3 1

Summing up, out of every 16 individuals we should expect:

Characteristic and class.	Relative frequency.	Per cent.
Median and split comb (a) + (c)...............	9	56.25
Median and unsplit comb (b) + (d)............	3	18.75
No-median and split comb (e).................	3	} 25.00
No-median and no-split comb (f)..............	1	

* For further discussion of the Y-shaped comb see page 65.

12   INHERITANCE IN POULTRY.

On account of the difficulty of deciding in the case of any young chicks whether 'no-median and no-split comb' is present, the last two classes are combined in the right-hand column of this table and in columns α and β of the table above.

In comparing the fit of the expected percentages on the two hypotheses with the actual, it is seen that hypothesis α is the better. However, the real test will come in the F_3 generation. On hypothesis α the single-combed individuals bred together should produce only median-combed offspring. On hypothesis β they should produce some without median comb.*

First generation hybrids crossed with Minorca. When the Y-shaped comb is crossed back on the single comb of the Minorca (No. 26 ♂) the following distribution of comb characters was obtained:

Characteristic.	Frequency.	Per cent.
Single comb	21	52.5
Cleft comb	†19	47.5
Papillæ or "absent"	0	0.0

This result accords well with the expectation that 50 per cent of the combs shall be of the pure Minorca type and 50 per cent of the heterozygous type.

2. NOSTRILS.—*First generation.* No case occurred of a typical high nostril; this characteristic is dominated by the narrow nostril; but this dominance is imperfect. In three cases (5 per cent) the nostril is recorded as one-half high, *i. e.*, having an aperture one-half as high as the extreme. In the other cases, placed in class 2, the breadth was less, but still evidently influenced by the germinal representative of the "high" characteristic. In two skulls that were dissected the processus nasalis of the premaxillary bone was present.

Class.	Characteristic.	First generation.		Class.	Characteristic.	Second generation.	
		f.	Per cent.			*f.*	Per cent.
1	Narrow	47	83	1	Narrow	45	52.9
2 {	a. One-third high...... 7 b. One-half high...... 3	10	17	2 {	a. One-third high......16 b. One-half high...... 6	22	25.9
3	High	0	0	3	High or nearly high.....	18	21.2
	Total	57	100		Total	85	100.0

Second generation. In the second generation the high nostril appears again in full or nearly full size in 21 per cent of the cases. Class 3 includes two

---

*This topic is discussed again, generally, at page 68.

† Including one thick comb with a median ridge in addition to the cleft comb, forming a typical pea comb. Seen in unhatched chick XVIII, 103.

cases in which the records read three-fourths high; but even in the Polish fowl the high nostril varies somewhat. Theoretically, we should expect 25 per cent of the second generation to have the recessive high nostril. The deficiency may indicate merely insufficient numbers, or perhaps some of the "one-half high" group truly belong in "class 3." The essential facts are, first, that high and narrow are segregated, and that in the second generation also dominance is frequently imperfect.

First generation hybrid crossed with Minorca. The heterozygous mothers all had narrow nostrils, as of course had the Minorca father. No true case of high nostril occurred. The recorded distribution is as follows:

Class.	Nostril.	No of individuals.
1	Narrow...	12
2	One-third to one-half high.......................	13

It seems probable that the 12 individuals with narrow nostrils belong to homogametous birds and the 13 individuals of class 2 to heterogametous birds.

3. CEREBRAL HERNIA.—*First hybrid generation.* Every bird was without a typical cerebral hernia. Nevertheless some of them showed clear traces of their mixed ancestry. On the frontal bone of all fowl is the so-called frontal eminence which is covered by fascia and the thickened skin of the crest. The profile of the skull from the apex of this eminence to the parietal is nearly straight, except for a slight concavity limiting the frontal eminence. In hybrids this concavity is frequently still more marked, the eminence being higher. Indeed, in one case (No. 405 ♀) the concavity is so marked that, as far as one can judge from the living bird, there is a slight hernia. We see, then, that though plain head is dominant it is incompletely so.

Second hybrid generation. The following is the distribution of this characteristic when the hybrids are bred *inter se*. Expectation is: 75 per cent without hernia, 25 per cent with hernia. The result agrees closely with expectation.

Characteristic.	Frequency.	Per cent.
Hernia absent...........................	75	76.5
Hernia present..........................	23	23.5
Total...............................	98	100.0

First hybrid generation crossed with Minorca. Since the first generation is DR and Minorca is D, half of the offspring will be pure dominants and half heterozygous, both halves showing the dominant "absence of hernia." The result agrees with expectation.

Characteristic.	Frequency.	Per cent.
Hernia absent................................	34	100
Hernia present...............................	0	...
Total...........................	34	...

4. CREST.—*First hybrid generation.* The crest is present in every hybrid old enough to show a crest, yet always in reduced size. Crest is dominant, but the dominance is imperfect. The crest is larger in the females (fig. 5) than in the males (fig. 6).

Characteristic.	Frequency.	Per cent.
Crest absent................................	0	0
Crest present...............................	70	100
Total...........................	70	...

Second hybrid generation. All records, from embryo chicks as well as adults, give:

Characteristic.	Frequency.	Per cent.
Crest absent................................	23	*30.7
Crest present...............................	52	69.3
Total...........................	75	100.0

* Fig. 11.

Expectation is that crest will be absent in 25 per cent of the cases. The excess is probably due to the fact that, since crest develops late, it was noted as absent when it would have appeared in adult life. To test this I give a table based on hatched chicks only:

Characteristic.	Frequency.	Per cent.
Crest absent................................	11	21.2
Crest present...............................	41	78.8
Total...........................	52	100.0

This gives a close approximation to expectation, with a slight excess of *crested* individuals. The exact statistical proportion, with its possible errors of classification, is less essential than the fact of reappearance in about one-fourth of the cases of the recessive characteristic.

First hybrid generation crossed with Minorca. Since plain-head is R, the cross is of the order DR × R; from which we should expect an equal number of heterozygous (crested) and pure recessive (plain-headed) offspring. The result, though based on small numbers, accords with expectation.

Characteristic.	All records.	Hatched chicks only.
Crest absent (RR)........	12	6
Crest present (DR)......	8	6
Total...........................	20	12

5. COLOR OF TOP OF HEAD.—*First hybrid generation.* All records give:

Characteristic.	Frequency.	Per cent.
Wholly black..	36	64.3
Black and white.	20	35.7
Total	56	100.0

It is to be noted, first, that the white of the crest tends to disappear in the later molts, some birds which showed it at 2 months losing it by 6 months, or showing white at the tip only of the crest feathers. Further, with two exceptions, all crests with white feathers belong to females (which have larger crests than males). The two exceptional males are still young and have only a trace of white remaining; this will probably disappear in the next molt. Third, the proportion of white to black in the crest is always small—much smaller than in the Polish crest. The result looks like an imperfect dominance of black.

Second hybrid generation. Hatched chicks only give:

Characteristic.	Frequency.	Per cent.
Wholly black.............................	24	47.1
Black and white.	27	52.9
Total.	51	100.0

I interpret this irregular result to be due to the imperfect dominance of black. Twenty-five per cent of the individuals have wholly black gametes and 25 per cent wholly black-and-white, or mosaic, gametes. The 50 per cent with mixed gametes tend to be black, but contain white in varying proportions. Something more than one-fourth of the black-and-white headed individuals are males.

First hybrid crossed with Minorca. All records give:

Characteristic.	Frequency.	Per cent.
Wholly black	21	91.3
Black and white	2	8.7
Total	23	100.0

Expectation, assuming complete dominance of black, is 100 per cent of black individuals. Result shows incomplete dominance. Of the black-and-white headed individuals, one is a female; the other died too early for the sex to be determined. Here, again, dominance is less perfect in the female.

6. CORRELATION OF CHARACTERS.—In the Single-comb Black Minorca and in the White-crested Black Polish there is an assemblage of characters that are nearly always associated in those races. The first hybrids have another constant association unlike either of the parents, viz., split comb, black crest (in the male) without cerebral hernia, and low to medium nostrils (fig. 6). In the second generation of hybrids, on the other hand, occur combinations of characters both of the parental species and also of the first generation of hybrids. These combinations are of the most varied sort, so that characteristics always found associated in one parent species may here be found dissociated. When hybrids are bred *inter se* the following combinations are obtained:

No.	Comb.	Crest.	Hernia.	Nostril.	Number of cases. Actual.	Number of cases. Calculated.
1	Y	Present	Absent	Narrow	17	15
2	Y	Present	Absent	High	0	5
3	Y	Absent	Absent	Narrow	8	5
4	Y	Absent	Absent	High	0	1.6
5	Y	Present	Present	Narrow	1	5
6	Y	Present	Present	High	0	1.6
7	Y	Absent	Present	Narrow	3	1.6
8	Y	Absent	Present	High	0	0.5
9	I	Present	Absent	Narrow	8	7.3
10	I	Present	Absent	High	0	2.5
11	I	Absent	Absent	Narrow	12	2.5
12	I	Absent	Absent	High	0	0.8
13	I	Present	Present	Narrow	2	2.5
14	I	Present	Present	High	0	0.8
15	I	Absent	Present	Narrow	0	0.8
16	I	Absent	Present	High	0	0.3
17	o o	Present	Absent	Narrow	1	7.3
18	o o	Present	Absent	High	8	2.8
19	o o	Absent	Absent	Narrow	0	2.5
20	o o	Absent	Absent	High	1	0.8
21	o o	Present	Present	Narrow	4	2.5
22	o o	Present	Present	High	5	0.8
23	o o	Absent	Present	Narrow	0	0.8
24	o o	Absent	Present	High	0	0.0

This table gives the distribution of characteristics in 70 individuals. Grouping the individuals under certain alternative characters, we have the following relations of actual and calculated frequency of occurrence of each characteristic:

Characteristic.	Actual.	Expected.	Characteristic.	Actual.	Expected.
Split comb........	29	35.0	Hernia present....	16	17.5
Single comb.......	22	17.5	Hernia absent.....	54	52.5
Papillæ...........	19	17.5	Nostril high.......	14	17.5
Crest present......	49	52.5	Nostril low........	56	52.5
Crest absent......	21	17.5			

The actual never deviates far from the expected.

If, however, we compare the actual number of cases of each of the combinations with the calculated the result is instructive. For example, in the absence of correlation of characters we should expect a high nostril to be associated with a single comb in 5 or 6 of the 22 cases; but it is never found so associated. In fact *a high nostril never occurs in this cross dissociated from a rudimentary comb*. On the other hand, it appears that a low nostril may be associated with a rudimentary comb, but in unexpectedly few cases, 4 instead of about 14. Two of the 4 records are from embryos in the shell, in which therefore adult characteristics were not fully developed and the other two cases are recorded as one-half high. It is quite possible that an atypical nostril and absence of true comb are always associated (fig. 11).

In order not to complicate the table too much, the correlation between crest and color of the crest feathers was omitted. A subsidiary table is consequently given here:

		$f.$
Crest present...................	{ Black............................	18
	{ White and black	23
Crest absent....................	{ Black............................	9
	{ White and black.............	6
		56

Whether crest is present or absent white occurs on the head; but it is more apt than not to occur when the crest is present and less apt than not when the crest is absent.

CONCLUSIONS.

In the cross under consideration no characteristic is inherited in strictly Mendelian fashion, for in no case is dominance complete. The nearest approach to typical Mendelian dominance is exhibited, in the present cross, only when crest is crossed with no crest. The new additive characteristic—crest—

is dominant. But the crest of the first generation hybrids is always of small size. Likewise, plain head is dominant over cerebral hernia, but some of the hybrids have exceptionally high frontal prominences. The white color of crest is recessive in the male hybrids, but is not entirely shut out from the females. The high nostril is recessive, yet the presence of its representative in the hybrid gives the latter abnormally wide nostrils. Finally, the comb affords us a case of an organ in which neither parental form can be said to be dominant without such an extension of the term as to render it quite vague. Every individual shows a modified comb—the Y or O shaped comb. This is a new form—a heterozygous form—that probably reappears in the heterozygotes of each generation.

The facts of correlation show that crest is by no means dependent on cerebral hernia. At the same time I doubt if the absence of present correlation disproves the hypothesis that the crest was the result of the hernia. It is at least conceivable that a characteristic that arose as a response to the stimulus of an abnormal ontogenesis should become hereditary and independent of the stimulus. As for white color on the top of the head, it is dissociable from the crest, for wholly black-crested second hybrids occur.

Series II.—Single-comb White Leghorn and Houdan.

STATEMENT OF PROBLEM.

This cross was undertaken for comparison with that between Minorca and Polish, and to test the inheritance of plumage color, extra toe, and face feathering.

THE RACES AS A WHOLE.

The Leghorn (fig. 15) is typical of the Mediterranean class of poultry—slender, tall-legged, close-feathered, nervous, and non-broody—the same class as that to which the Minorca belongs. The white Leghorns came originally from northern Italy.* They have been bred in America since 1834. The single-comb variety is one of the most widely bred of our races and has the reputation of being the greatest egg-producer. Considering its white plumage, its transparent skin, with a trace of yellow, and its red iris, it comes very near to being an albino race, but the retina is pigmented.

The Houdan (fig. 16) comes from France. It, like the Dorking, has doubtless descended from the 5-toed fowls of the Romans, described by Columella, which they probably carried to Gaul in their conquest of that

---

* Wright, 1902, p. 385; Wyckoff, 1904, p. 788.

country. This may have been crossed with "the old crested race of Caux."* The Houdan may be regarded as one of the fundamental types.

TABLE OF CHARACTERISTICS.

No.	Characteristic.	Single-comb White Leghorn.	Houdan.
1	General plumage color	White	Black, white-tipped.
2	Color of upper mandible	Yellow	Light horn.
3	Nostrils	Narrow	High.
4	Comb	Single (rarely cleft behind, No. 11.)	2-pronged or V.
5	Face feathering	Plain	Whiskered.
6	Chin feathering	Plain	Bearded.
7	Dorsal head plumage	Plain	Crested.
8	Dorsum of cranium	Plain	Domed.
9	Foot color	Yellow	White.
10	Number of toes	Four	Five.

DISCUSSION OF CHARACTERISTICS.

1. GENERAL PLUMAGE COLOR.—In the Leghorn this is typically white, and the most highly selected birds are without trace of black specks or yellowish lacing. The yellow lacing is hard to get rid of. The Houdan color consists typically of black feathers occasionally tipped with white (fig. 16).

2. COLOR OF UPPER MANDIBLE.—The clear yellow of the mandible of the white Leghorn is part of the general pigmentation of the skin. Much yellow pigment is deposited over the body. It shows prominently in the tarsal scutes. The Houdan mandible is clear black.

3. NOSTRILS.—The high nostrils of the Houdan (fig. 12) are like those of the Polish fowl (page 7).

4. COMB.—The comb of the Houdan in America is the so-called V-comb. It differs from the Polish comb (page 7) in that the two horns arise from the sides of a median swelling (Fig. B). In England the Houdan is cultivated with a leaf comb consisting of two broad, flat expansions of the horns arising from a median ridge like "a butterfly with open wings."† It thus resembles the posterior part of a Y-comb (fig. 8). The single comb of the Leghorn is very large and lops in the female to the right or left side of the head.

FIG. B.—Dorsal view of beak of Houdan 9A♂ showing pair of clublike papillæ, c, that represent the V-comb, c. f., culminal fold.

*Petersen, C. E., 1905, p. 961, quoting P. Megnin: "Élévage et engraissement des volailles."
†Hurst, C. C., 1905, p. 132.

5. WHISKERS, OR MUFF.—This is a bunch of long feathers growing from the sub-orbital and post-orbital region of the head. This characteristic, of whose origin nothing is known, has been engrafted on several of the other French breeds: the Crêvecœur, the Faverolle, the Du Mante, the Cossack, the Bourbourg, etc. The muff occurs also on breeds which have little in common with the French fowl, *e. g.*, the Sultan and the Orlaff and Pavaloff of Russia.

6. BEARD.—This consists of a number of long feathers growing from the middle of the chin and upper throat region (fig. 16). There is a fold of skin here from which the feathers arise. Such a beard is usually associated with the muff. The fold of skin, "dewlap," is found in some Indian Games and, less marked, in the Dark Brahma male.

7, 8. CREST; CEREBRAL HERNIA.—These are indistinguishable from those of the Polish (pp. 8–10).

9. FOOT COLOR.—The brilliant yellow color of the tarsus of the Leghorn is strikingly different from the colorless or dirty gray tarsus of the Houdan. There must be a special yellow pigment in the skin of the former which is absent in the latter.

10. NUMBER OF TOES.—The possession of an extra toe (fig. 13) is an ancient characteristic of poultry. The Latin author Columella, speaking of the fowl kept by the Romans, says: "Those hens are reckoned of the purest breed which are 5-clawed, but so placed that no cross-spurs arise from the legs." Since the tendency to extra toes must have arisen spontaneously once, there is always a possibility that it has done so several times, and it is by no means certain, although probable, that the extra toe of the Polish is genetically connected with that of the Roman fowl referred to. The following record of occurrence of extra toes in poultry is interesting, since in this case no relation to the Roman fowl is probable. A writer* in Der Zoologische Garten states that Carl Bock in his "Reich des weissen Elefant," p. 267, relates that he, in a journey from Tschengmai, on the third day reached Muang Hawt, a way station on the road to Mulmen. This village is distinguished for its 6-toed fowl. Again, the Silky fowl, which certainly have little in common with the Dorking, have extra toes (page 59). The extra toe is to be regarded as a sport which has appeared at different times and which is easily maintained as a racial characteristic. The question of the inheritance of such a sport is always interesting. The Houdan has typically only one extra toe, making 5 in all; it is occasionally found with 6. Bateson and Saunders (1902, p. 98) sometimes got 6 toes in hybrids between Leghorns and Dorkings. The length of the extra toe and the completeness of bifurcation are very variable.

---

* Langkavel, B. 1886, p. 35.

PREVIOUS INVESTIGATIONS.

During the progress of my experiments appeared the second report to the Evolution Committee of the Royal Society by Bateson and others. This contains a paper by Hurst (1905, p. 133) giving his results with White Leghorn male × Houdan female. These will be considered in comparison with my results.

MATERIAL.

The *mothers* were two Houdan hens* (fig. 16) purchased from a dealer as pure stock. They agreed well with the standard requirements. When bred with a Houdan male they produced only typical Houdans.

The *father* was a Single-comb White Leghorn† likewise of unknown ancestry. The plumage of No. 13A is devoid of black pigment, and mated with White Leghorn hens it has produced only White Leghorn offspring.

RESULTS.

1. GENERAL PLUMAGE COLOR.—*First hybrid generation.* Of 41 individuals all were white in plumage (fig. 17), but almost without exception both in down plumage and that of the adult there were traces of black on one or more feathers, particularly those of the back; more especially was this true of the females than of the males.

Hurst (1905, p. 133) got 11 black chicks out of 105 offspring and in the first plumage these developed into 6 black (all pullets) and 5 barred (all cockerels). Here also females have more pigment than males. Of the white chicks all except two developed black ticking. Doubtless these two were males.

Second hybrid generation. When these hybrids were crossed *inter se*, out of 50 individuals 9 were markedly pigmented like the Houdan. This is 18 per cent of all cases, expectation being 25 per cent. With larger numbers Hurst (1905, p. 138) got 24.3 per cent black. Equally striking is the occurrence of many *pure white* individuals along with the impure whites. The pure whites that were reared to maturity proved to be males; the impure whites were females.

First hybrid (No. 87 ♂) crossed with white Leghorn (No. 71 ♀). The father was pure white; the mother was speckled with black. Of 22 offspring all were white, but some had single pigmented feathers. All males (9) are pure white, excepting No. 562, which has some black on two feathers of the left wing coverts, and No. 649, which has one-half of one right wing covert black. My only certain female has a score of partly black feathers. Hurst (1905, p. 139) obtained 66 clear white chicks and 69 white, ticked with black. I

---

* Nos. 8 and 11, received July 1, 1904, from Geo. C. Ely.
† No. 13A, received Sept. 15, 1904, from a farmer.

judge this equality to indicate a difference of color in sex; or else the pigmented individuals are heterozygotes. Possibly the females are the heterozygous forms—the males homozygous.

2. COLOR OF UPPER MANDIBLE.—This assumes its final condition so late in life that definite statistics will not be given now.

First hybrid generation. A few young are recorded as showing yellow. The rest are white; this is probably the young condition of the light horn of the adult Houdan. Light horn seems dominant.

Second hybrid generation. A few cases of black mandible are recorded, even in the young, where pigment is less developed.

First hybrid (87) crossed with white Leghorn. All older chicks have white mandibles; there are no blacks.

3. NOSTRILS.—*First hybrid generation.* Of 25 individuals, all but one have a nostril of one-half width or less—thus approaching the white Leghorn type. The exceptional individual (No. 386 ♀) has a typical high nostril and is peculiar in respect to comb also. Only one individual is recorded as having as narrow a nostril as the Minorca.

Second hybrid generation. Forty-nine individuals give:

Characteristic.	Frequency.	Per cent.
Narrow (24) and intermediate (8)	32	65.5
High	17	34.5
Total	49	100.0

On the assumption that "narrow and intermediate" includes pure-narrow and heterozygous individuals, while "high" includes recessive, pure-high individuals, we should expect 75 per cent and 25 per cent in the two classes respectively—only an approximate agreement with the actual.

First hybrid (No. 87 ♂) crossed with white Leghorn (71 ♀). The father has a "one-fourth" nostril; the mother, of course, a typical "narrow" one. Of 24 individuals 12 are recorded as narrow; 12 as intermediate of some grade. This gives the ratio 1:1, which we expect, assuming the intermediate nostril to be the heterozygous type; the narrow, the pure type.

4. COMB.—*First hybrid generation.* Of 41 individuals 40 have the Y-shaped comb in some form (fig. 17). This comb resembles that of the Minorca × Polish hybrid. There is no case of a single comb in this generation, but there are two cases in which the posterior end of the comb is merely much thickened. On the whole the present series of cleft combs differs from the former in that a smaller proportion of the comb is split—no cases of wholly split or cup combs occur, although in one important case (87 ♂) two-thirds

Self Reliance Books

Get more historic titles on animal and stock breeding, gardening and old fashioned skills by visiting us at:

http://selfreliancebooks.blogspot.com/

of the comb is cleft. A new characteristic of this series of cleft combs is the occasional appearance of a *median* comb lying between the two wings of the cleft comb—a posteriad continuation of the single part of the comb. This condition appears in three cases (258 ♂, 259 ♂, 448 ♂). It is important for the interpretation of the cleft comb. It gives the posterior part of the hybrid comb the triple condition characteristic of English Houdans.

The one case that lacks the Y-shaped comb is No. 386 ♀ (with high nostrils). She has only a pair of papillæ. Hurst (1905, p. 133) got no single comb in 105 offspring.

Second hybrid generation. Fifty-five individuals show the following distribution of comb forms:

Characteristic.	Frequency.	Per cent.	
		Actual.	Expected.
			α β
Single	17	30.9	25 (18.75)
Y comb.....	23	41.8	50 (56.25)
o o or absent.............	15	27.3	25 (25.00)
Total................	55	100.0	100 (100.00)

The Y comb being the heterozygous form should appear in 50 per cent of the cases; each of the other forms in 25 per cent. The deviation from expectation is of the same character as in Series I. That the approximation to theory is less close is probably due to the smaller total number. Hurst (1905, p. 138) obtained 56 ordinary single combs in 226, or 24.8 per cent.

First hybrid (87 ♂) crossed with white Leghorn (71 ♀). The Y-shaped comb crossed with single gives, in 26 individuals:

Characteristic.	*f*.	Per cent.
Single..................................	15	57.7
Cleft, etc. (*see* Remarks)................	11	42.3
Total........................... ..	26	100.0

REMARKS: Including two cases in which a median ridge runs through the cleft comb. Of these one is a *nearly typical pea comb* except that the side lobes are higher than the median one. Including, furthermore, one case of an arrow-shaped comb, two parallel V's occurring in front and behind, respectively, being joined by a median ridge. Including, finally, two cases of cup-comb.

Here the expected equality is approached. Hurst (1905, p. 139) obtained 60 ordinary single combs in 135 individuals of this cross, or 44.4 per cent.

5. FACE FEATHERING.—*First hybrid generation.* Of 24 recorded cases all show the muff (fig. 17).

Second hybrid generation. All individuals (26) whose face feathering was observed are recorded as muffed; concerning a greater number (35) the record is silent. What has become of the expected 25 per cent of muffless individuals? It is possible, but on strict chance hardly probable, that the muffless individuals all died young. A decisive answer to our question must await further experimentation.

First hybrid (87 ♂) crossed with white Leghorn (71 ♀). Only one parent is muffed. Muffed and non-muffed offspring occur in approximate equality; but even in the adult muffing is not full in amount. This cross confirms the result of the first that muffing is dominant, but it is not perfectly so.

6. BEARD.—In the *first hybrid generation* all individuals are bearded. When these hybrids are crossed with the white Leghorn about half of the offspring are beardless. Beard is dominant.

7. CEREBRAL HERNIA.—In the *first hybrid generation* all (24) individuals were without external evidence of the cerebral hernia. In the *second hybrid generation* out of 45 individuals 11 had the hernia (fig. 14) and 34 had none, or 24.4 per cent and 75.6 per cent respectively. When the *hybrid was crossed with the white Leghorn* (71 ♀) no individual with the hernia appeared. The cerebral hernia is a recessive characteristic. However, the height of the frontal dome is variable, even in the pure-bred Houdans, and on at least one occasion the cerebral prominence in an unhatched hybrid was so high that it was doubtful whether or not it might represent a hernia. Here, as in Series I, plain-headedness, though clearly dominant, is imperfectly so.

8. CREST.—*First hybrid generation.* Of 25 individuals all are crested. Hurst (1905, p. 134) gets the same result. *Second hybrid generation.* Of 19 individuals 6 are non-crested, or 31.6 per cent, approaching the expected 25 per cent. The remainder are crested. *First hybrid (87 ♂) crossed with white Leghorn (71 ♀).* Of 15 individuals 6, or 40 per cent, are without crest. Expectation is 50 per cent. Crest is clearly dominant, yet in the first hybrid it is never so large as in the Houdan. This fact is, I take it, due to imperfect dominance. It may, however, be associated physiologically with the absence of a cerebral hernia.

9. FOOT COLOR.—In the *first hybrid generation* this always becomes white in the adult, although sometimes yellow in young birds. In the *second generation of hybrids* bred *inter se* or with the White Leghorn stock, yellow reappears as leg color. Statistics would be misleading on this point, as permanent leg color is reached only on maturity. It may be concluded that white is dominant over yellow.

10. NUMBER OF TOES.—*First hybrid generation*. Among 37 individuals the following distribution appears:

Number of toes.	f.	Per cent.
4-4	6	16.2
4-5	8 ⎫ 31	83.8
5-5	23 ⎭	
Total	37	100.0

Hurst (1905, p. 133) got among 105 birds 103 with trace of extra toe (including duplication of nail and hyperphalangia) and two without any such trace. The difference in the proportions of extra and normal toes between Hurst's and my results is partly a matter of classification and perhaps partly due to the real difference in potency of the extra-toe characteristic in the two strains.

Second hybrid generation. To learn if the individuals with 4-4 toes were merely imperfect dominants or true recessives I mated two of them (Nos. 84 ♀ and 86 ♀) with their brother, a 4-toed cock (No. 83 ♂). Of 23 offspring, 17 were normal-toed and 6 had extra toes on one or both feet, or nearly 25 per cent with extra toes. Expectation, on the other hand, was either (*a*) if the 4-4-toed were recessives there should be no extra toes, or else (*b*) if extra toe here merely fails to dominate there should be 75 per cent with extra toes. Hurst (1905, p. 150) mated together two 4-4-toed hybrids and got 14 extra-toed to 8 normal, or 63.6 per cent extra-toed. He also mated the same 4-4 hybrid cock with a 4-toed Hamburg Cochin hen, and about half the offspring had extra toes. He concludes: "These results prove that the apparently recessive feet with no trace of extra toe . . . . are in reality DR's, as both birds gave chicks with e. t. when bred together and with pure recessives." I am inclined to doubt if this is the whole story, for one of my two 4-4-toed hybrid hens, namely, No. 86, mated with 83 ♂, gave 12 offspring with 4-4 toes and no certain offspring without normal toes. This looks as though 86 ♀ and 83 ♂ were both truly recessive. No. 84 ♀, on the other hand, produces extra-toed and normal-toed individuals in about equal proportions. Further experiments with 83, 84, and 86 are planned for 1906.

Second hybrid (87 ♂) mated with white Leghorn (71 ♀). The father (87) has an extra toe on the left side only. Of 25 offspring 17 have 4 toes on each side, 6 have 5 toes on each side, and 2 have an extra toe on one side only.

11. CORRELATION OF CHARACTERS.—The first hybrids between white Leghorns and Houdans show a fairly constant association of characteristics, a white plumage flecked with black, white mandible, narrowish nostrils, Y-shaped comb, muffed and bearded face, reduced crest on a domeless head,

white legs, and toes that are usually but not always meristically abnormal. When the hybrids are bred *inter se* we get varied combinations of characteristics, as follows (D meaning dominant and R, recessive):

Comb.	Nostril.	Hernia.	Toes.	Plumage.	Actual No. of cases.
Y	Low (D)	Absent (D)	Normal (R)	White (D)	6
	Low	Absent	Normal	Black and white	1
	Low	Absent	Extra	White	3
	Low	Present	Normal		0
	Low	Present	Extra		1
	High	Absent	Normal	White	1
	High	Absent	Normal	Black and white	1
	High	Absent	Extra		0
	High	Present			0
I	Low	Absent	Normal	White	6
	Low	Absent	Normal	Black and white	0
	Low	Absent	Extra	White	3
	Low	Absent	Extra	Black and white	0
	Low	Present	Normal		*2
	Low	Present	Extra		0
	High	Absent	Normal		0
	High	Absent	Extra		0
	High	Present	Normal		0
	High	Present	Extra		0
• ○	Low	Absent	Normal	White	0
	Low	Absent	Normal	Black and white	0
	Low	Absent	Extra		0
	Low	Present	Normal		1
	Low	Present	Extra		0
	High	Absent	Normal	White	4
	High	Absent	Normal	Black and white	1
	High	Absent	Extra	White	3
	High	Absent	Extra	Black and white	1
	High	Present	Normal	White	3
	High	Present	Normal	Black and white	1
	High	Present	Extra		0
					38

*Fig. 14.

CONCLUSIONS.

In the series of crosses between the White Leghorn and Houdan, Mendelian results were obtained as in the first series. Dominance, however, is frequently imperfect. The plumage color of the offspring of a pure homogametous white-and-black and a white are rarely pure white. Likewise in the second hybrid generation impure whites occur. Also, nostril-height exhibits imperfect dominance of the narrow type. The muff, beard, and crest, though always present in the first hybrid generation, are apparently always reduced. The cerebral hernia, though recessive, affects the dominant normal skull.

A *heterozygous form* results from hybridizing the single and V-shaped comb. The cleft comb is a neomorph, of which more will be said in the sequel (page 65).

Polydactylism does not readily fall into the Mendelian formula. Hurst's results, although suggestive, need support from other experiments.

Few characters are correlated; the second hybrid generation exhibits all combinations, except that high nostril and single comb do not occur together here any more than they do in the Minorca-Polish hybrids.

Among heterozygous individuals, with Y-shaped comb, the combination of dominant characteristics (narrow nostril, no hernia, and white plumage) is the commonest, forming nine-thirteenths of all. These are also, apart from the Y-comb, all Leghorn characteristics. It appears, too, with the given parentage, that *normal* toes are usually present. The low nostril and cerebral hernia combination occurs several times with or without extra toe. The combination refutes the notion of Wright (quoted at page 8) that there is any necessary relation between cerebral hernia and "cavernous" nostril. A high nostril was in two instances (both of which died very young, one before hatching) associated with a Y-comb, but it is doubtful if the median portion would have developed.

The single comb may occur associated with a hernia (*e. g.*, No. 443, fig. 14), with extra toes and with mottled plumage, but in my records so far it is never associated with a high nostril.

The V-shaped comb is commonly associated with high nostril, but rarely with a low one, despite the fact that low nostril is dominant. It occurs on white individuals thrice as frequently as on black-and-white ones; it shows no preference for the extra toe.

The hernia is never found dissociated from the crest; but the crest occurs three times as often as the hernia.

Series III.—Houdan and Single-comb Black Minorca.

STATEMENT OF PROBLEM.

This series was undertaken to compare the behavior of the Houdan with that of the Polish (Series I, page 6) when crossed with the Minorca.

THE RACES AS A WHOLE.

The Houdan is described at page 18 and the Black Minorca at page 6. Both of the races are fundamental and old. The Houdan contains the larger assemblage of new characteristics.

TABLE OF CHARACTERISTICS.

Characteristic.	Houdan.	Discussed at page—	Single-comb Black Minorca.	Discussed at page—
1. General plumage color..	Black and white..	19	Black..........	29
2. Comb form............	V...........	19	I................	7
3. Nostril form..........	Wide	19	Narrow........	7
4. Crest.................	Present........	20	Absent.........	
5. Cerebral hernia........	Present........	20	Absent.........	
6. Muff.................	Present........	19	Absent	
7. Beard...............	Present........	20	Absent	
8. Foot color...........	White..........	20	Black..........	29
9. Toes................	5-5...........	20	4-4............	

MATERIAL.

As *mothers*, Nos. 8 and 11 (page 21), original Houdan stock, and later their daughters, Nos. 60 and 81, were used. No trap nests were employed in this series and consequently mothers are not exactly known.

Father: No. 27, bred at the station, August, 1904, son of No. 12, Minorca cock (page 6) and a Minorca hen.

RESULTS.

1. GENERAL PLUMAGE COLOR.—The young hybrids, like the young black Minorcas, contain much white on the belly and primaries. In later moltings the white is replaced by black, but even at five months the primaries are sometimes mealy or white-spangled. Except for this the hybrids have lost the Houdan white and are of the Minorca color. Minorca uniform black is dominant over the Houdan mottling.

2. COMB.—*First hybrid generation.* Of 20 offspring 18 have a Y-shaped comb like the hybrids of Polish and Minorcas. In two cases (of egg embryos) the comb was recorded as single, but this is probably only the limiting condition of the Y-shaped comb, which is here also the heterozygous form.

3. NOSTRIL FORM.—*First hybrid generation.* In no case does the nostril width exceed one-half. As in Series I and II, there is imperfect dominance of the narrow form. However, the nostril in this cross runs lower than in the Leghorn × Houdan cross.

4. CREST.—This is present in all first hybrids reared to a sufficient age to render an opinion possible.

5. CEREBRAL HERNIA.—F_1. Always absent. One egg embryo, which died at about 17 days of incubation, had a vesicle protruding, uncovered by skin, from the top of the head at the usual position of the hernia. Such a teretological case is not uncommon in straight-bred Houdans. It is noteworthy to find it here where the cerebral hernia is recessive.

6, 7. MUFF AND BEARD.—F_1. Present in all older hybrids.

8. FOOT COLOR.—F_1. Always black as in the Minorca.

9. TOES.—Of 21 hybrids, 12 have 5–5 toes, 3 have 5–4 toes, and 6 have 4–4 toes. The proportion without extra toes is higher than in the Leghorn × Houdan first cross, being there 16.2 per cent, here 28.6 per cent.

CONCLUSIONS.

The following characteristics apparently exhibit alternative inheritance: Plumage color, nostril form, crest, cerebral hernia, muff and beard, leg color, and number of toes. Dominant are: Uniform black plumage color (imperfect), narrow nostril (imperfect), crest (imperfect), cerebral hernia (imperfect), muff and beard (imperfect), and black leg color. Of these the crest, muffling, and black leg color are positive characters in de Vries' sense and are dominant. The color pattern of the Houdan yields here to black as it does in Series II to white; mottling is recessive to solid color. The comb, crest, muffling, and extra toe are inherited essentially as in Series II. Striking is the nearly universal imperfection of dominance.

Series IV.—Single-comb White Leghorn and Rose-comb Black Minorca.

STATEMENT OF PROBLEM.

This cross was undertaken to learn the inheritance in these races of the characteristics described below.

THE RACES AS A WHOLE.

The Leghorn has been described at page 18; the Minorca at page 6.

TABLE OF CHARACTERISTICS.

No.	Characteristic.	Single-comb White Leghorn.	Discussed at page—	Rose-comb Black Minorca.	Discussed at page—
1	General plumage color..	White..........	19	Black..........	29
2	Comb form.............	Single..........	19	Rose........	29
3	Foot color.............	Yellow..........	20	Blue-black....	29

REMARKS ON THE CHARACTERISTICS.

1. GENERAL PLUMAGE COLOR.—Black is one of the constituents of the color of *Gallus bankiva*, being the chief color of the breast. Just how a wholly black condition of plumage was attained is of course not exactly known; there are, however, many instances known of melanic sports among birds. It is probable that wholly black varieties have arisen as a result of excessive production of the black pigment, melanin.

2. COMB FORM.—The rose comb is a broad mass of erectile tissue replacing the single comb. Anteriorly it overhangs the nostrils and extends over and back of the eyes. The upper surface is covered by numerous tubercles. These do not, in young birds and females, run wholly at random but tend to lie in five or more parallel lines. Posteriorly the rose comb ends in a finger-like process or spike. The rose comb is the last term of the series starting at the single comb and having the triple or pea comb for its middle term.

3. FOOT COLOR.—The dense black color of the scutes of the foot of the Minorca constitutes a positive or additive characteristic as contrasted with the pigmentless condition of the yellow-footed White Leghorn.

MATERIAL.

As *mothers* three White Leghorns were used, of unknown ancestry but reputed pure. They had fair Leghorn points except as noted in the descriptions given below. When mated with White Leghorn 13A ♂ they produced only white offspring.

10A. Feathers all white; comb strictly single.

11A. Feathers mostly white, but some are sooty. The single comb is cleft behind for about one-fourth of its total length.

12A. Feathers all white; comb strictly single.

The *father*, No. 9A, also of unknown parentage, has a large rose comb 90 mm. long by 44 mm. broad at its widest part. The tubercles are very irregular, but five rows of them can be discerned. The plumage is prevailingly black, but many feathers of the back are tipped with white and several primaries are almost or quite white.

RESULTS.

This series of experiments has been carried as yet through only the first hybrid generation.

1. PLUMAGE COLOR.—Eighty-three first hybrids were obtained of which 74 were white, either pure or with some black feathers, and 9 were deeply pigmented. Females A and C yielded only white offspring. Female B, on the other hand, produced chiefly dark birds, recorded as "blue" or "black-and-white." She was the mother of the 9 pigmented birds just mentioned. B's germ cells are probably mixed. The only two of B's offspring reared to maturity are blue like the so-called "Andalusian breed" (fig. 54, pl. XVII). Now blue is a combination of black and white and is a "heterozygous form." If blue birds, one of which is male, the other female, breed together, both pure black and impure white, as well as blues again, are to be expected in the proportions of 1:1:2 respectively.* Of the white offspring of both A and C it is noteworthy that the males are mostly pure white (*i. e.*, without trace of black, although often suffused with yellow), while the females are always specked with black.

2. COMB FORM.—Of 80 first hybrids 40 have single comb and 40 rose comb. This result indicates first that the cock is a heterozygote and consequently produces in its germ glands two kinds of germ cells, viz, those with the single-comb and those with the rose-comb determinants and second, that rose comb is dominant. Then: R single × DR rose gives 50 per cent DR rose and 50 per cent RR single.

Bateson and Saunders (1902, pp. 102, 103) find rose comb of Wyandotte or White Dorking dominant over single comb of the Leghorn. Hurst (1905, p. 134) crossed a White Leghorn with a Black Hamburgh (rose comb)

---

*Tegetmeier (1867, p. 185) states that blue Polish bred together throw cuckoo, white or speckled produce. Wright (1902, pp. 399–401) states that "Andalusians" constantly throw black and also white chicks. Blue chicks are frequently produced by crossing black and white. Wright (1902) mentions such a result from crossing black and white Langshans (p. 291) and Wyandottes (p. 318). Such blues also throw whites and blacks. Inheritance of blue is discussed by Bateson and Saunders (1902, pp. 131–132) and by Bateson and Punnett (1905, pp. 118–119). In the latter paper it is stated that of 75 offspring of Andalusians 17 were "white splashed, 36 blues, 22 blacks." A blue bred to a white produced 34 blue and 20 white splashed; bred to a black it gave 27 blue and 19 black.

and got rose combs in all of the offspring. Here, too, rose comb is dominant over single comb.

Rose comb is a positive variation. It behaves in Mendelian fashion. Although a neomorph, it is dominant.

3. FOOT COLOR.—Excluding from consideration all but the older hybrids, 40 showed foot coloring as follows:

Black, bluish, willow, or green	9
White	7
Yellow	24
Total	40

This result indicates that one of the parents (probably the male) is a heterozygote containing traces of some white-legged ancestor. Yellow appears to be dominant over white and black, but to be imperfectly so. The black × yellow gives green or willow; the whitened yellow is "white." Hurst (1905, p. 137) finds that when blue-footed and yellow footed individuals are crossed, the offspring have light-blue feet. Yellow is in his case recessive.

CONCLUSIONS.

In general, plumage color, foot color, and comb form are inherited in Mendelian fashion. White plumage is dominant, although imperfectly so; wherefore we have spotted whites and even blues. Rose comb is dominant; whether perfectly so can not be determined until later. Yellow foot color seems to be dominant, but is imperfectly so, even the yellow legs showing traces of black.

Series V.—Single-comb Black Minorca and Dark Brahma.

STATEMENT OF PROBLEM.

This cross was made to see the result of uniting two races as unlike as possible in origin and general form.

THE RACES AS A WHOLE.

The Minorcas have been already described at page 6. The Dark Brahma race was originally imported from India near the home of *Gallus bankiva*; yet it differs from it as much as does the Minorca. It is a blocky, short, stout-legged bird, is fluffy in plumage, and has a small pea comb and small wattles. It is, moreover, sexually dimorphic. The male (fig. 19, pl. v) has much more black in its plumage and is very differently marked from the penciled or barred female (fig. 18). The Dark Brahma has so many characters unlike those of the Jungle fowl that it is now thought to be chiefly derived from a different ancestor, namely, that of the Aseel and Indian races.

TABLE OF CHARACTERISTICS.

No.	Characteristic.	Single-comb Black Minorca.	Dark Brahma.
1	General color.................	Black in both sexes..	Complex black, red and white pattern; sexes dissimilar.
2	Wing coverts (wing bar).......	Black	Black, white, and red.
3	Comb.....	Single.............	Pea.
4	Earlobe color.................	White, red mottling.	Red.
5	Iris color....................	Brown	Yellow.
6	Foot and beak color..........	Blue-black.........	Yellow.
7	Foot feathering...............	Absent	Present and heavy.
8	Vulture hock..	Absent	Present.
9	Wing coverts.................	Black	Black, white, and red.

REMARKS ON THE CHARACTERISTICS.

1. GENERAL PLUMAGE COLOR.—The adult Minorca has a completely glossy black plumage. In the chick of two weeks the whole ventral surface is covered with a white down, and certain of the primaries, especially those at the distal end of the series, are partly or wholly white. The coverts overlying these reminges are usually white also. The white on the wing gradually disappears in successive molts.

The adult Dark Brahma has a sexual dimorphism of color. The female (fig. 18) is nearly uniformly penciled black and straw color. The hackles have a broad whitish margin and the inner half of the vane is solid black. The male (fig. 19) has solid glossy greenish-black feathers in the tail, white feathers on top of the head, in the middle of the back and upper wing coverts; below black. Feathers with narrow black central stripe and broad white margin (lacing) constitute nape, hackle, and saddle. Red occurs on wing bar and humeral patch. In the down plumage Brahmas of both sexes are longitudinally striped buff and black.

2. WING BARS.—The wing bar of the male Dark Brahma depends upon the fact that the wing coverts of the second and third rows (which are black at the base) have a white tip and a transverse band of red in the middle between black and white. In the higher coverts and on the shoulder the red still occurs, but it is reduced in extent.

3. COMB.—The single comb is found on *Gallus bankiva* and has sometimes been regarded as the only ancestral form. The pea comb is a distinct type, consisting of a median comb bordered on each side by an accessory comb. The origin of the pea comb is obscure but certainly ancient. Wright (1902, pp. 265, 330, 339) contends that it arose in the Aseel, a type of oriental fowl regarded as the ancestor of the Malays and Indian Games and believed not to have originated from *Gallus bankiva*, but to represent a distinct species. According to this view the pea comb has not arisen from the single, but is coördinate with it and of equal age. From the ancestral breed it has become

fixed upon others. Although not typical of the Malay breed (which has a small "walnut" comb), it often appears when two walnut combs are bred together.* The pea comb is found also in certain derived races, particularly in the Brahma, and from this it has been engrafted on various other breeds, notably on the Plymouth Rock of America and on the so-called "Buckeyes." †

4. EARLOBE COLOR.—Two main color types of earlobes are found in poultry, red and white. The former occurs in the Jungle fowl, Aseels, Indian Games, Javas, Dorkings, Cochins, Brahmas, and other foundation stock; consequently it must be regarded as the more primitive. The white earlobe seems to have arisen in the Mediterranean type. It finds its highest expression in the White-faced Black Spanish. It has become a constituent of the Houdans and La Flêche among the French breeds and of the Hamburghs.

5. IRIS COLOR.—Among poultry this ranges from a gray or pearl through yellow to orange, red and bright red on the one hand, or through a dirty red or bay to brown and black on the other. The red type seems to belong to the descendants of the Jungle fowl; it is found in the Jungle fowl, in most Games, in most Mediterranean breeds, in most French fowl, in the Dorkings, and in many of their derivatives. The lighter colors, yellow and pearl, are found in the Aseels, Malays, Indian Games, often in the Brahmas, attesting their origin from the Aseel group, also in many Cochins, where it is said to be "very hereditary" (Wright, 1902, p. 320). The dark colors—hazel, brown, and black—are found in certain Game fowl of dark plumage, the Brown-Red Games, the Birchen Games and the Black Sumatras. They are found also in the Black Javas of America, in the Langshans, and sometimes the Orpingtons. Both the Langshans and Orpingtons have derived their eye color from the Java. Dark-brown eyes are found among the Mediterranean fowl only in the Black Minorca which we have to do with here. Whence acquired by the Minorca is uncertain; possibly indirectly from the Java. Finally, a perfectly black iris is found in the Silkies, where it is probably due to the melanic sport that has made also skin and connective tissue black.

6. FOOT COLOR.—This varies with the general skin color. The primitive color of the *Gallus bankiva* group is an olive, commonly called "willow." This is found in ordinary Game fowl. The Aseel-Malay-Indian type has yellow feet. To this type belong the yellow feet of the Brahma and the Cochin and doubtless also of the American Dominiques, which have transmitted it to the Plymouth Rocks and Wyandottes. Finally, the Leghorns have bright yellow feet.

By increase of the black pigment in certain birds of dark plumage there have been produced from the willow foot the blue, blue-black, and leaden blue

---

* Wright, 1902, p. 325.

† American Standard of Perfection, 1905, p. 79.

feet of numerous races, *e. g.*, Black Cochins, Black Wyandottes, Black Java, Black Leghorn, Black Minorca, White-faced Black Spanish, Hamburgh, the French fowl, black and dark-colored Games, and the Silky. By decrease of pigment are derived the white feet of the Dorkings and Houdans. This loss of pigment may be regarded as a mutation. It is associated with red or yellow eyes.

Considering the Aseel type and the *bankiva* type as specifically distinct, the cross of the yellow foot and the blue-black foot in the present series is a cross between specific characteristics.

7. FOOT FEATHERING.—In *Gallus bankiva* and in the Aseel-Malay group the feet are without boots. The same is true of the Game fowl, although minute feathers are sometimes found on their feet. Foot feathering is found among various species of birds; among scratching birds, in grouse, ptarmigans (*Tetreo, Bruasia, Lagopus*), among some pigeons, and the owls. Typically absent from the Gallinæ, it has cropped out in the Brahma, Cochin, and, probably independently, in the Silky and Sultan. In these groups it has been preserved because of its importance in brooding or because it has struck the fancier's eye.

8. VULTURE HOCK.—This consists of long stiff quill feathers projecting backward at the heel joint. It is found among poultry only in the Cochin-Brahma group and its derivatives. This characteristic is a good example of a progressive variation.

MATERIAL.

The cock used in this cross, No. 122 (fig. 19), was a bantam Dark Brahma, weighing 1,900 grams, received (February, 1905) from F. H. Hodges,* Red Bank, New Jersey, who is a successful breeder of this variety. The cock is typical of his kind.

The hens were four Single-comb Black Minorcas, Nos. 14 (fig. 3), 16, 18, and 28, of which the three former were purchased of Mr. George C. Ely in July, 1904, and No. 28 was hatched at the station in August, 1904, from one of the purchased hens mated with the full-blooded Minorca cock No. 12.

RESULTS.

Only the first generation of hybrids has been reared up to the time of writing.

1. GENERAL PLUMAGE COLOR.—In all cases (41) the hybrids are prevailingly black. Usually the feathers of the occiput and nape are laced with white, much more in the males than in the females, and the hackles of the male are obscurely barred or splashed with white (fig. 21). Evidently the white lacing of the Dark Brahma is trying to assert itself. The color of the wing coverts is considered in the next paragraph. The down of the young

---

* Foot marked "F. H. H., 164."

is dead-black, being without the longitudinal stripes of the Dark Brahma young, and, for the most part, without the white wing feathers and ventral aspect of the young Minorca. Nevertheless, exceptionally, one finds the chin and part of the throat of the young white, the head feathers may be tipped with white, and in one or two instances a little white occurs on the wing. The young plumage seems to be a neomorph, but on the whole it belongs rather to the Minorca type than to the more primitive Game type of juvenile coloration.

2. WING COVERTS.—In 14 grown male hybrids of which I have records, a more or less prominent wing bar, formed by black, red, and straw-colored feathers in the third or fourth row of wing coverts, occurs (figs. 20, 21). The five females are wholly black, but even in these the wing coverts are barred with an iridescent purple black; consequently a disturbance of the coloration of the wing coverts may be said to be typical of the hybrids. The wing bar of the Dark Brahma male dominates over the black wing of the Minorca, but it dominates imperfectly.

3. COMB.—In all cases the pea comb of the Brahma dominates over the single comb. Critical examination shows, however, that the pea comb of the hybrid is not always typical. Frequently the whole structure, and especially the median ridge, is abnormally high (fig. 21), and, on the other hand, in a few cases the lateral ridges are hard to make out. The dominance is imperfect.*

4. EARLOBE COLOR.—The earlobe in every case contains both white and red. The result is not a blend, however, but is particulate, the white appearing at the center. As red is rarely wholly absent from the Minorca's earlobes, whereas white is wholly absent from that of the Dark Brahma, it may be possible to bring inheritance of earlobe color under the general formula and speak of the white earlobe as being imperfectly dominant.

5. IRIS COLOR.—The iris of the hybrid is yellow, rarely with a trace of red or reddish brown. The type of the Dark Brahma is dominant, but imperfectly so.

6. BEAK AND FOOT COLOR.—This is always black in the hybrid. However, the black is rarely the blue-black of the Minorca, but it is usually a brownish black frequently tinged with yellow, particularly on the under side of the toes. Black is imperfectly dominant.

7. FOOT FEATHERING.—In all cases the hybrids have feathering on the feet. In many cases this is clearly reduced in amount from what is found in the Dark Brahma. Foot feathering is imperfectly dominant (fig. 20).

8. VULTURE HOCK.—This is absent in every case, although about a quarter of the cases show the feathers of the heel much larger and more removed

---

*The inheritance of the pea comb of the Dark Brahma has not been considered in the recent studies of others. The pea comb of the Indian Game is found by Bateson and Saunders (1902, p. 94) to be imperfectly dominant over the single comb of the White Leghorn.

from the foot than in the Minorca. Plain feathered heel is dominant, but not perfectly so.

CONCLUSIONS.

This series of experiments is only begun. Conclusions as to dominance are tentative until tested in the second hybrid generation. The Minorca characteristics appear to dominate in (1) general black color, (4) white ear-lobes, (6) black foot and beak, and (8) absence of vulture hock. Dark Brahma characteristics appear to dominate in (2) wing bar, (3) pea comb, (5) yellow iris, and (7) foot feathering. In every case, however, dominance is imperfect. In some cases, at any rate, (5, 7), it is the new, additional, or positive characteristic that dominates.

Series VI.—White Leghorn and Dark Brahma.

STATEMENT OF PROBLEM.

THE RACES AS A WHOLE.

It is proposed to investigate the behavior of characteristics when the heavy-bodied, short and stout legged, loose-feathered, dark-colored Asiatic type is crossed with the slender, tall-legged, close-feathered, white Mediterranean type. Both types are ancient, but the Brahma must be regarded as nearer its ancestral form, Aseel-Malay-Indian, than the Leghorn is to the Jungle fowl.

TABLE OF CHARACTERISTICS.

No.	Characteristic.	White Leghorn.	Dark Brahma, female.	Dark Brahma, male.
1	Hackle color.....	White	Straw, black-penciled...	Black, straw-laced.
2	Wing bar........	White........ ..	Buff, black-penciled ...	Black, red, and white.
3	Wing bow........	White	Buff, black-penciled .	Black and white.
4	Tail color........	White	Black, straw-penciled...	Greenish black.
5	Comb	Single (page 19).	Pea (see page 32).	
6	Earlobe..........	White, red-edged (page 33).	Red (see page 33).	
7	Iris color	Red (page 33) ...	Yellow (see page 33).	
8	Vulture hock.....	Absent	Present (see page 34).	
9	Foot feathering..	Absent	Present (see page 34).	
10	General form....	Narrow, slender..	Broad, chunky.	

REMARKS ON THE CHARACTERISTICS.

1. HACKLE COLOR.—Among most poultry that are of broken color the hackle feathers are unlike those of the rest of the plumage. They have a dark center and a lighter lacing. In the Malays and Indians they have a red center edged with green. In the *Gallus bankiva* female the hackles have a black center (with straw-colored shafting) and straw-colored lacing. This is the type of hackle feather found in the male Dark Brahma. It is found,

in both sexes, among many other breeds. The hackle of the female Dark Brahma (fig. 18) differs from that of the male in that the broad black center is barred, or penciled, with straw color.

3. WING BOW.—The wing *bar* is described for the male at page 32. In the male Dark Brahma the feathers of the fourth and higher rows of wing coverts have their distal halves white forming the wing bow. No such distinct wing bar and wing bow occur in the female, the feathers of this region being uniformly penciled like the others.

MATERIAL.

There are two sets of experiments in this series. In the *first set* the mothers were White Leghorn Bantams, Nos. 127 and 128, probably heterozygotes with black, further discussed at page 39. The paternal Dark Brahma, No. 122, has been already referred to at page 34. In the *second set* the mother was Dark Brahma, No. 121. She has the same history as No. 122. She is a prettily penciled bird (fig. 18). The father was White Leghorn Bantam, No. 126, described at page 39.

RESULTS.

Of the first set 19 offspring are recorded, including 8 in the shell. Females 8, males 5. Of the second set 27 are recorded, including 8 in the shell. Females 14, males 4.

1. GENERAL PLUMAGE COLOR.—The result differs in the two sets and the two sexes, and it is otherwise variable.

Plumage color.	Females.			Males.			Unknown sex.			Total.
	First set.	Second set.	Total.	First set.	Second set.	Total.	First set.	Second set.	Total.	
Nearly pure white	4	3	7	3	2	5	7	11	18	30
White + much black and red pigment as barring	0	0	0	2	0	2		2	2	4
Like Dark Brahma female	4	3	*7	0	0	0				7
Barred	0	3	3	0	2	2				5
Black or nearly so	0	5	5	0	0	0				5

The results are explicable on the hypothesis that all of the White Leghorn Bantams, Nos. 126 ♂, 127 ♀, and 128 ♀, contain white gametes and also gametes bearing red pigment, black pigment, and the barred pattern.

3. WING COLORATION.—In the *first set* the wing coloration is like that of the plumage in general, except that in the females marked like the Dark Brahma the coverts contain much red (fig. 23).

Second set. Of 7 prevailingly white hybrids three show red or buff on the wings;† of the 5 black-and-white (penciled) birds all but one show red or purple on the wings; of two adult black hybrids one shows buff. Three other

---

* Fig. 23.   † Fig. 25, Plate VIII.

females are marked like the Dark Brahma female. Red pigment is commoner in this set, with White Leghorn father, than in the first set, with Dark Brahma father. This speaks for the hypothesis that red has come from the White Leghorn, as, according to usual experience, the father tends to determine coloration.

4. TAIL COLOR.—*First set.* Of 11 offspring, 8 have a white tail, the prevailing color of the body; in one case the tail is white except for one black feather, and in two cases it, like the body in general, agrees with the Dark Brahma female in being black with buff penciling (fig. 23).

Second set. Of 15 hybrids, 6 are nearly or wholly white on the tail, one has two black feathers, 5 are black, two are black-and-white barred, and one is black with buff, as in the Dark Brahma female. The tail color tends to resemble that of the general body.

5. COMB FORM.—In all cases of adult hybrids of either set, the comb is pea (fig. 24). Pea comb is consequently here also dominant over single comb.

6. EARLOBE.—Both the Brahma solid red and the White Leghorn white, red-margined earlobes appear in about equal numbers. It is probable that my heterozygous White Leghorn bantams have been early crossed with some red-lobed race.

7. IRIS COLOR.—This is definitely established only in mature birds. All eyes show more red than the Dark Brahma and the tendency is to redden with age; consequently red is probably dominant.

8. VULTURE HOCK.—This is absent in all cases (fig. 22). One hybrid has the hock feathers a little elongated. Short feathering at the heel is dominant.

9. FOOT FEATHERING.—*First set.* Of 19 hybrids having the Dark Brahma father, 3 unhatched chicks are recorded as non-booted. Of the remainder, 8 are slightly or very slightly booted. Three adults have a medium covering of feathers on the foot. The Brahma tendency toward booting has been diluted by the cross with the Leghorn.

Second set. Of 24 offspring of Dark Brahma mother, all have well-developed boots. This constitutes a striking case of a *difference in reciprocal crosses*. Booting is probably here, as elsewhere, dominant, but frequently very imperfectly so.

CONCLUSIONS.

Of the nine characteristics, the following exhibit clear alternative inheritance, the dominant characteristic being printed in italics:

Pea comb vs. single comb.
No vulture hock vs. vulture hock.
Booted foot vs. unbooted (when Brahma is mother).

The other characteristics can not for one reason or another be so easily classified. The red of the wing bar seems to behave like a unit character and is independent of the coloration of the rest of the body.

The inheritance of booting is peculiar in that in the first set, Leghorn mother and Brahma father, the booting fails to show that clear dominance which is almost universal; yet I can hardly suspect the purity of the Dark Brahma male. It would seem that in this series also the mother transmits booting more strongly than the father.

Series VII.—Black Cochin Bantam and White Leghorn Bantam.

STATEMENT OF PROBLEM.

This experiment was undertaken to learn the method of inheritance where one parent is pure white and the other pure black.

THE RACES AS A WHOLE.

The Black Cochin Bantam, also called Black Pekin, is a diminutive of the Cochin (fig. 26). It is stated by Wright (1902, p. 499) that the Pekins came in 1860 from the city of that name. The original color was buff; the black has probably been engrafted on the race by a cross with some small black race. The Cochins are closely allied to the Brahmas and share with them a chunky form, stout and densely feathered feet and red face and earlobes. The White Leghorn has been discussed at page 18.

TABLE OF CHARACTERISTICS.

No.	Characteristic.	Black Cochin Bantam.	White Leghorn Bantam.
1	General plumage color.	Black.........	White.
2	Earlobe color..........	Red............	White, with trace of red.
3	Vulture hock.........	Present.........	Absent.
4	Foot feathering........	Present.........	Absent.

REMARKS ON THE CHARACTERISTICS.

1. GENERAL PLUMAGE COLOR.—In the Black Cochin this is a deep greenish black. No trace of white appears anywhere.

2. EARLOBE COLOR.—In the Black Cochin this is of the dark red or bay characteristic of all the Aseel-Malay group.

3. VULTURE HOCK.—This is well developed in the Black Cochin (see page 34).

MATERIAL.

The *mothers* were four Black Cochin Bantams,* very similar, each heavily booted and weighing about 600 grams apiece. Trap nests were not used, but owing to special peculiarities the eggs of the separate mothers were distinguished as A, B, C, and D.

The *father* was a White Leghorn Bantam, No. 126, purchased January, 1905, from the Cyphers Incubator Company, together with two hens (Nos.

---

* Nos. 129, 130, 131, 132, received February, 1905, from Mr. H. B. Kutschbach (fig. 26, pl. IX).

127, 128). Mated with the hens, nine young were produced. Four of these were typical White Leghorns without black; three others were white except that black feathers occasionally appeared. One (No. 213) was nearly solid black and one (No. 229) was black with nearly every feather barred with white. It is plain that the strain I have has not been wholly purified of black. This is indicated also by the fact that No. 128 has every feather peppered with black—a heterozygous form of coloration.

RESULTS.

1. GENERAL PLUMAGE COLOR.—Of 26 hybrids, 11 were pure white or had only a little black; 7 were black, sometimes with a little white, and 8 were barred black and white (fig. 27). The results confirm the view that White Leghorn Bantan No. 126 ♂ has black germ cells. The barred condition is unexpected and is probably recessive in the White Leghorns.

2. EARLOBE COLOR.—In all cases (10) of adults but two, the earlobe is red; in the remaining two some white is mixed with the red. The red earlobe is probably dominant, but imperfectly so.

3. VULTURE HOCK.—Out of 13 cases 11 have clearly no vulture hock and two show a slight enlargement of the heel feathers. Vulture hock is probably recessive.

4. FOOT FEATHERING.—Every hybrid is booted, but the booting is less heavy than in the Dark Brahma (fig. 27). Booting may be dominant, but it is not perfectly so.

CONCLUSIONS.

The male parent is heterozygous and probably contains at least three sorts of gametes, viz, pure black, pure white, and barred, the last being a mosaic.* The black of the mother is recessive to all of these. The occurrence of barred mosaic is of interest, but it is of unknown origin. The ancestral red ear-color and the new "booting" are both dominant. Dominance is, however, imperfect.

Series VIII.—White Leghorn Bantam and Buff Cochin Bantam.

STATEMENT OF PROBLEM.

This series was undertaken to determine the method of inheritance of buff when combined with a white plumage coloration.

THE RACES AS A WHOLE.

The White Leghorn Bantam has been described at page 39. The Buff Cochin Bantam (fig. 28) is a diminutive Buff Cochin, which resembles in form the Black Cochin (p. 39). Cochins as we know them to-day (the name was formerly applied to a different, now extinct, race) seem to have been imported into this country and also into England from eastern China

---

* Castle and Allen, 1903, p. 606.

in the year 1847. The earliest importations were buff. According to McGrew (1904, p. 526):

> In many of these retreats, *mi-aus* or monasteries, thousands of specimens of Buff and Partridge China (Cochin) fowls are annually raised, and in other places the colors are mixed. The Kinkee (gold flower) colored birds are the most esteemed, both as regards antiquity and purity. . . . Hoangho is the oldest [of the] *mi-aus*, and its records show that this same race of fowls was cultivated by the brotherhood 1,500 years ago.
>
> Buff and Partridge Cochins are indigenous to the temperate and more southerly portions of the empire. This is corroborated by naturalists and travelers. Mr. Gabb, the well-known English naturalist, says: "According to my view, a black or white Cochin is an improbability, if not an impossibility, as a natural product of a tropical or subtropical region. The natural color of the feathers of the poultry in the zone of Cochin China would be buff or yellow, or some of the varieties of these colors, but never black or white, except by accidental variation."

Other testimony presented by the same author is of the same sort and establishes the fact that Buff Cochins are a primitive, foundation race of great antiquity.

TABLE OF CHARACTERISTICS.

No.	Characteristic.	White Leghorn.	Discussed at page—	Buff Cochin.	Discussed at page—
1	General color	White	19	Buff	41
2	Earlobe color	White	33	Red	33
3	Vulture hock	Absent		Present	34
4	Foot feathering	Absent		Present	34

REMARKS ON THE CHARACTERISTICS.

1. GENERAL PLUMAGE COLOR.—The buff color of the Cochin is, as has been shown above, of high antiquity. From the Buff Cochin it has been transferred to many other breeds by crossing. Thus there are Buff Wyandottes, of which McGrew says (1901, p. 24): "Two distinct lines were produced under different methods. One was formed from Wyandotte-Buff Cochin cross; the other came through the Rhode Island Red-Wyandotte cross." The Rhode Island Red is, however, as is well known, a direct descendant of the Buff Cochin. The Buff Plymouth Rocks were derived directly or indirectly from the Buff Cochin (McGrew, 1901, p. 25). The history of the Buff Leghorn is the same—the offspring of a yellow Danish Leghorn cock and Buff Cochin pullets mated with a yellow Leghorn hen. The produce, three-fourths Yellow Leghorn and one-fourth Buff Cochin, gave* "the first Buff Leghorns ever shown." The Buff "Orpingtons"— a highly modern and mongrel breed—have a similar history, being chiefly Buff Cochin and Dorking (Wright, 1902, p. 296).

The origin of the buff as it occurs in the Cochins can only be guessed at; but there are important facts to be considered. First, it appears that the buff color is very inconstant even in China. Says a traveler: "No two can

---

*Wyckoff, 1904, p. 527.

be found of exactly the same color; some are a chestnut color, others darker, and some quite light" (McGrew, 1901, p. 527). Of the Buff Cochins as first imported to England, Wright (1902, p. 245) says: "The buff colors were much subdivided, ranging from the lightest silver buff and silver cinnamons through lemons and buffs to the deep colored cinnamons which would now be called almost red. Originally, also, the birds were not uniformly buff over the whole body; even prize-winners were such as would now be called 'tricolored,' the breast being lemon or orange buff, the hackles and saddle much darker, and the wing darker still, even a red." From all of this it is plain that buff is only a diluted form of red—a color that is abundant in the plumage of the Malay and Indian breeds, and the replacement of all black by this buff is probably due, originally, to a xanthic "sport."

MATERIAL.

The *mother* was the White Leghorn Bantam No. 128, a heterogametous bird, already discussed at page 40. The *father* was a Buff Cochin Bantam, No. 545 (fig. 28), original stock, of whose ancestry nothing is known.

RESULTS.

1. GENERAL PLUMAGE COLOR.—Thirty-one offspring show the following distribution of color: White, 9; white and buff, 9; white and black, 4; white, black, and buff, 2; black and buff, 4; black (all juvenile), 3.

Calling the germ cells of the mother equally white and white-and-black and regarding the buff as (imperfectly) recessive when paired with white, we have—

Characteristic.	$f.$	Percentages.	
		Actual.	Expectation.
White (and buff).	18	58.1	50
White-and-black (and buff).	13	41.9	50
Total.	31	100.0	100

Of the white and buff heterozygotes, white only appears in 9; the remainder show some buff. White is dominant, but imperfectly so.*

Wright (1902, p. 244) states in regard to crosses between white and buff Cochins that in the early days they "bred most amazingly in regard to color. . . . From one brood of ten chickens of this cross two pullets were pure black; two pullets and three cockerels black with more or less gold in the hackles, and marked wings; the other three darkly penciled birds."

Hurst (1905, p. 134) finds that crosses between White Leghorn female and Buff Cochin male (essentially the same crosses as mine) gave 60 chicks—

---

* But see fuller discussion of the heterozygous nature of my White Leghorns, page 40.

" 53 whites and 7 buffs. Of these 2 were apparently clear whites and 51 white patched with buff and brown, chiefly on the head, neck, and breast (18 of these were also oddly black-ticked); the 7 buffs were all more or less patched white." Hurst concludes that the white plumage color of the Leghorn is dominant over the Cochin buff, but that this dominance is incomplete in the majority of cases. He adds: "In F_1 the dominance of both white and black over buff is much less complete than that of white over black."

2. EARLOBE COLOR.—In all recorded cases the hybrids have a red earlobe, sometimes with a lighter colored, even yellowish, center.

3. VULTURE HOCK.—This is always absent in the hybrids. However, two cases show an elongation of the heel feathers.

4. FOOT FEATHERING.—In all cases the "boot" of the hybrid was reduced as compared with the Cochin parent. In 3 cases out of 31 no trace of feathers could be detected on the tarsus.

CONCLUSIONS.

Buff is recessive toward white, but the dominance of white is very imperfect, so that we may have various degrees of buffness in the hybrids. Black, or the mosaic black-white, appears to dominate over buff, but here again the dominance is frequently imperfect. Imperfect dominance is not revealed by a blending, but by sprinkling of the red pigment.

The *earlobe color* of the Cochins (Aseel type) dominates over that of the Leghorn, but not perfectly. *Vulture hock* is recessive, but not always perfectly so. *Foot feathering* may be said to be imperfectly dominant. But this case is of special interest because the result is practically a blend. Hurst (1905, p. 134) similarly states that out of 60 chicks from his cross, all had feathered "shanks," "but in every case the length or number of feathers was reduced to about one-half."

Series IX.—Tosa Fowl (Yokohama) and White Cochin Bantam.

STATEMENT OF PROBLEM.

This series of experiments was undertaken in the first instance to test the inheritance of the long-tailed characteristic of the Japanese long-tailed fowl (variously called Tosa fowl, Yokohama, Phœnix fowl, Japanese Game Shinowara-to, etc.).

THE RACES AS A WHOLE.

The Tosa fowl* (figs. 29, 31) has long been bred in Japan and plays a

---

* Professor Mitsukuri on the occasion of a recent visit to the Station for Experimental Evolution informed me that in Japan these birds are known as Tosa fowl, since they were originally bred in the province of that name, particularly at Shinowara. He further remarked that the feudal chief, or daimio, of that province had as his emblem or insignia a spear with a long cock's feather on it, and he made the interesting suggestion that the activity of the fanciers had been stimulated, not only by their satisfaction in long-tailed birds, but also by the desire of meeting the ever-increasing ideals of their chief as to the length of the feather of his insignia.

prominent part in Japanese art. On the authority of Chamberlain (1900), "as great a length of tail as 18 feet has been reached in the tail feathers, but even 12 feet is a rarity. From 7 to 8 or 11 feet is the usual length." Aside from the tail, the fowl has remarkably long hackle and saddle feathers of a golden color. Otherwise it closely approaches the European black-breasted Red Game, having, like it, retained most of the coloration of *Gallus bankiva*.

The Cochin fowl was used in the mating because its tail feathers are notoriously short and consequently afford a strongly opposed allelomorph.

TABLE OF CHARACTERISTICS.

No.	Characteristic.	Tosa fowl.	White Cochin Bantam.
1	General color............	Dark............	White.
2	Tail.......................	Long...........	Short.
3	Foot feathering..........	Absent..........	Present.
4	Foot color................	Willow..........	White.

REMARKS ON THE CHARACTERISTICS.

1. GENERAL PLUMAGE COLOR.—The colors of the male Tosa fowl* are very striking. The head is black; the feathers of the nape and the hackles are black proximately, but the exposed portion is red, becoming a deep mahogany on the middle of the back. The long saddle feathers are green laced with mahogany. The tail feathers are solid greenish-black. The breast, belly, and under tail coverts are black. The remiges are black, edged exteriorly with red. The coverts are black tipped with mahogany in varying amount, but so as to produce a marked red wing bar. The female Tosa fowl (fig. 30) has a black head and nape and golden hackles. The feathering of the back and saddle and the wing coverts are black mossed with rusty and have a straw-colored shaft. The breast is strongly tinged with buff. The *White Cochin*, on the other hand, is pure white (fig. 32).

2. TAIL.—The question of the origin of the long tail is of great importance. Any light on this question would illuminate the problem of specific differentiation and the origin of specific characteristics in general.

Hypotheses. In accordance with current theories of specific differentiation we have to recognize that this characteristic *may* have arisen:

(1) As a mutation. As such it would be brought into the same category with frizzled feathers or the cerebral hernia of Polish fowl. Professor E. Ray Lankester has referred to the condition as a sport.

(2) As the result of selection. This would be the most popular explanation. Romanes (1901, p. 302, fig. 95) includes this case as one of a number of typical proofs of the efficiency of artificial selection. Weismann (1904, II, pp. 124, 326) states definitely that the long tail is due to selection. The

---

* Fig. 29, Plate X.

underlying assumption in both cases is that the selection has been of minute favorable fluctuations rather than the conservation of sports. This is a conceivable hypothesis.

(3) As a result of functional hypertrophy of the feather follicle. By artificial treatment the blood supply to the follicles might be stimulated so as to make the feather grow longer. Such an effect might be inherited or not, as could be determined by breeding. If the offspring of the long-tailed fowl have a long tail, though untreated, then we would have, on the third hypothesis, an inheritance of an acquired character. Cunningham (1903) believes that the Tosa fowl is a demonstration of such inheritance.

Growth of tail feathers. The study of the long tail of the Tosa fowl leads us to consider the whole matter of feather growth and of the anatomy of the tail. In early months of their life chicks are constantly losing old feathers and gaining new ones built on a larger scale to meet the needs of the enlarging body. Later, these feathers are all molted during one period in the autumn. During development, the tip of the feather is formed first and growth continues at the base, within a sheath, for a shorter or longer period, depending on the eventual size of the feather. The reason why some feathers, like the contour feathers, are short is because growth quickly ceases. The feathers of the hackle, saddle, and tail of the male are long because the growth period is in them prolonged. The sickle feathers of the Leghorn are still growing for three months after the molting period; consequently they attain a length of 300 to 400 mm. If the period of drying up of the growth sheath at the base of the sickle feathers could be delayed in the Leghorn for an entire year they would become each a meter long. The reason for the great length of the tail feathers of the Tosa fowl is that they do not cease growing. In this respect they resemble the long hair of Angora guinea pigs, rabbits and cats, and the head hair of man.

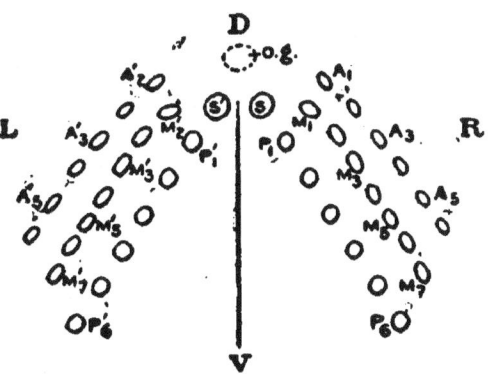

FIG. C.—Diagram of arrangement of the tail feathers. A_1-A_5, feathers of right anterior row; A'_1-A'_5, feathers of left anterior row; M_1-M_7 and M'_1-M'_7, feathers of right and left middle rows; P_1-P_6 and P'_1-P'_6, feathers of right and left posterior rows; o. g., oil gland; D-V, dorso-ventral line; R, right; L, left.

Morphology of the tail. As just intimated, only certain feathers of the tail of the Tosa fowl grow indefinitely. It is now necessary to describe the structure of the tail. The feathers have the following arrangement:

The posterior row (P) consists of broad feathers with rounded ends and constitutes the characteristic "fan" of the tail. The middle row (M) contributes the characteristic long growing feathers, those nearest the median line being longest. The sickle feathers (*S*) may belong to either row, so far as the adult position indicates; but, as growing feathers, they belong physiologically to the middle series. The anterior row (*A*) is, at the same time, the posterior row of tail coverts. The lateral feathers of this row are the smallest, owing to a late and brief growth. The long tail of the Tosa fowl is thus produced by the prolonged growth period of the middle row of feathers including the sickle, together with the more median feather of the anterior row.

Cause of excessive growth of tail; Cunningham's experiments. The cause of this prolonged growth of the median and sickle feathers is the crucial point. The latest student of the subject, Cunningham (1903, p. 232), quotes Mr. John Sparks as stating: "In order to ensure very great length of tail, the cocks ought to be kept on a perch . . . . . and the tail-feathers should be pulled gently every morning." Cunningham adds: "My own experiments . . . . . tend to show that this mechanical treatment of the feathers is the whole secret of the mystery." He describes in great detail how he stroked the tail of one of two cocks daily; the other not at all. When a feather stopped growing he pulled it out. He concludes (p. 248):

In the cock whose feathers were stimulated by pulling, growth did not go on at a more rapid rate, but continued for a longer time and produced a longer feather. Thus in cock A [not stroked] no growth took place after April 1, and the maximum length was 2 feet 4½ inches; while in cock B [stroked] growth continued till July 13, and the maximum length was 2 feet 9½ inches.

Half a page farther on Cunningham sums up thus:

The long-tailed cock in its perfection, therefore, is neither a sport nor a breed, but the product of artificial cultivation; and the excessive growth of the feathers is the result of stimulation applied to the individual. The more important part of the stimulation is not the mere pulling of the feathers, but the extraction of it which causes the growth of its successor.

One can not but remark that Cunningham here contradicts himself. After having laboriously pulled the feathers for over a year and found that the feathers are stimulated by pulling, he states: "The most important part of the stimulation is not the pulling but the extraction of the feather causing the growth of its successor." Does Cunningham indeed think that, originally, by extraction of a feather its follicle was so stimulated that it thereafter produced feathers which neither ceased to grow nor molted and, moreover, so affected the germ plasm as to produce a race with a tendency toward excessive growth of feathers? Certainly such a conclusion seems past belief.

Author's experiments. To see what influence, if any, stroking the tail feathers has upon their growth, I experimented upon two cocks. One (No. 3, "Admiral Togo") was stroked twice daily by passing the feathers of the middle and anterior row between the thumb and forefinger. The other (No. 7, "General Oyama") did not have its tail stroked. The two birds were treated similarly, except that Admiral Togo was confined to his perch during all but about two to six hours per day, while General Oyama had free run with the hens. The experiment was begun July 20, 1904, when the cocks were 103 days old, and was continued until March, 1905, when "Oyama" died of roup.

The relative growth of the corresponding feathers of the two males is shown in a series of curves (text figure D). The full line is the curve of

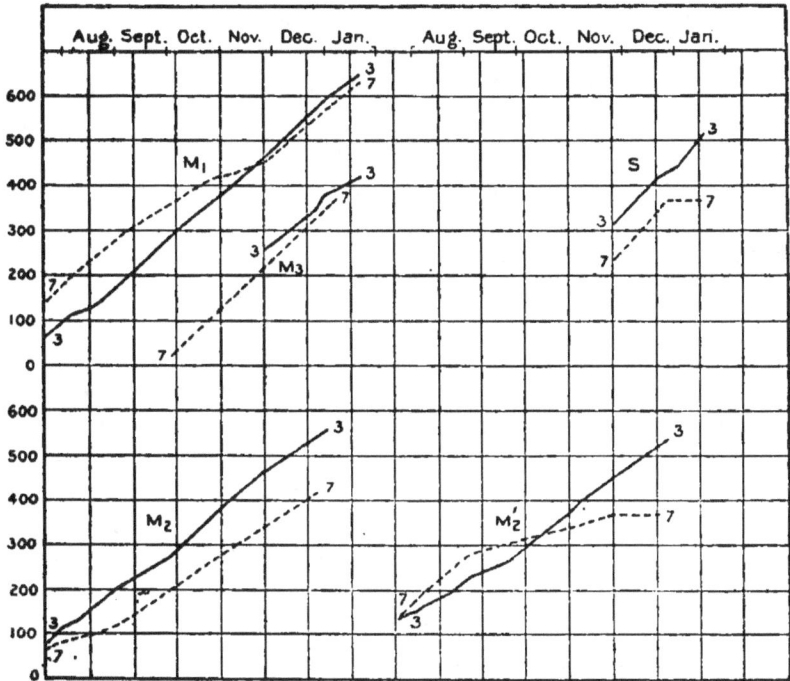

FIG. D.—Curves of growth of certain tail feathers of Togo (No. 3, full line) and Oyama (No. 7, dotted line). M_1, first right middle tail feather; M_2, M'_2, second right and left middle tail feathers; M_3, third right tail feather; S, right sickle.

the stroked feather; the dotted line that of the unstroked. These curves show several things.

First, the average rate of growth of one of the tail feathers in the Tosa fowl is about 3 mm. per day, or an inch a week. Consequently, if growth is uninterrupted and the feather does not break, it should come to be over a meter long by the end of one year. The extremely long feathers—5 meters or more—on record are acquired by (*a*) rapid growth, (*b*) continuous growth, (*c*) preservation of the tail from breakage, and (*d*) long life of the individual. If stroking has any effect it must be by altering one or more of these elements.

Second, the curves show fluctuations in the rate of growth due to fluctuations in the condition of the fowl.

Third, in the case of those feathers that were studied during the greatest period, viz, M_1, M_2, M'_2, the stroked feather grew more rapidly than the corresponding unstroked.

Fourth, the unstroked feathers of No. 7 ceased growing earlier than the corresponding stroked feathers of No. 3.

Whether the third and fourth items are due to differences in treatment or to other peculiarities of the fowl can not be asserted definitely. In any case the feathers all eventually, at about six months, ceased to grow. Owing to the death of No. 7 soon after, the comparison had to be abandoned. The result agrees with Cunningham's in that stroking prolongs the period of growth; but the result, depending on three feathers, can hardly be generalized. It would not be surprising, in view of what we know of functional hypertrophy, if it were some day demonstrated that stroking always prolongs the growing period of a feather. This is, however, an entirely different matter from proving that the origin of the long-tailed condition of the Tosa fowl was due to, and its maintenance in some way depends upon, stroking.

A few further experiments have been made with Admiral Togo (fig. 31). I have found, in confirmation of Cunningham, that if a feather that has ceased to grow be forcibly removed it is quickly replaced by another that continues to grow. Thus a feather pulled out January 1, 1905, had grown steadily to November 1; but as the bird was needed for breeding and could not be confined, the tail has repeatedly broken off. In September, 1905, it measured over 900 mm.

As a further criterion of the value of manipulation in causing this great growth of the tail feathers of the Tosa fowl, it becomes important to see how this physiological characteristic is inherited when crossed with a short-tailed individual. This consideration led to the present series of experiments.

The tail of the Cochin fowl is the shortest of all races of poultry. Thus Wright (1902, p. 245) expresses the ideal of the fancier: "The tail of the cock should be as short . . . . . as possible."

3. FOOT FEATHERING.—While the Cochin is very heavily feathered on the foot, the Tosa fowl is typically clean-legged. However, No. 3, which is not the father of any of my hybrids, shows a few bunches of rudimentary feathers or hairs on the tarsus.

4. FOOT COLOR.—The willow foot of the Tosa fowl is derived directly from the Jungle fowl. The white foot of the White Cochin seems to be an albinic form of the yellow foot derived from its Indian-Malay ancestry.

MATERIAL.

FIRST GENERATION.—The mother was a White Cochin Bantam, No. 35a (fig. 32), of unknown origin, but apparently pure in respect to the four characteristics here under consideration. The father was imported from Japan, having been purchased in New York city, January, 1904. It has a dark Game coloration (fig. 29).

SECOND GENERATION.—Two hybrid cocks, Nos. 53 (fig. 34) and 95 (fig. 35), were successively bred to their sisters, Nos. 58 (fig. 33), 94, 96, and 98.

RESULTS.

1. GENERAL PLUMAGE COLOR.—*First hybrid generation.* Of 7 offspring, 3 cocks and 3 hens developed their adult plumage. The males were all of the male Tosa-fowl coloration except that every feather was repeatedly barred with white (figs. 34, 35, 37A). The females were all of the female Tosa-fowl coloration except that the shafting was much broadened (fig. 37); also the saddle feathers and the proximal secondaries were obscurely barred black-and-buff.

Second hybrid generation. Among 57 individuals we have the following distribution of plumage color:

Color.	No.	Per cent.
White	16	28.1
Pigmented	41	71.9

The original white color has reappeared in about one-fourth of the cases (fig. 38); plumage color segregates in the germ cells of the first hybrid generation in true Mendelian fashion. Of the 16 whites, only 5 were without trace of reddish pigment. Such pigment occurred on the breast, top of head, and remiges. The purity of the germ cells from which these whites sprang—the completeness of segregation—is not always perfect.

The 41 pigmented individuals show a curiously mixed lot of coloration. Of 14 mature *females*, 6 are like the female Tosa fowl, without barring, but sometimes with wider shafting than male Tosa fowl. The remainder have feathers of the back and wing coverts barred with lighter, even with white—a condition not found in the female first hybrids. One of these (No. 659) shows a mixture of female Tosa and female *Partridge Cochin* coloration. As no Partridge Cochin is involved in the immediate ancestry, this looks like a "reversion;" the characteristic has probably lain latent in the White Cochin. Of 10 *males*, two show no trace of white, and may, consequently, be considered as homozygous. The remainder are more or less barred with white. One bird (No. 646) shows a remarkable mixture of Tosa and male Partridge Cochin coloration.

2. TAIL LENGTH.—*First hybrid generation.* All the three males reared developed abnormally long middle tail feathers. One of these birds died young. The second bird (No. 53, fig. 34) lived to be exactly one year old. Its sickles were 427 mm. long and had stopped growing. It had suffered a severe paralytic stroke four months before its death. The remaining cock (No. 95, fig. 35) had at 11½ months sickle feathers 360 mm. long and still growing. These feathers had thus grown at a rate of about 1 mm. a day, or

only one-third that of its father. The long-tailed characteristic of the male has been inherited, but in a reduced form.

Second hybrid generation. Still immature.

3. FOOT FEATHERING.—*First hybrid generation.* Of the 7 individuals all have the feet feathered ("booted") and the females are provided with a "vulture hock." The feathering is usually less than in the Cochin.

Second hybrid generation. Among the 55 individuals of this generation all degrees of foot feathering were obtained. Eight cases are recorded as "heavily booted," 27 as "booted," 13 as slightly booted, and 7 as non-booted. The classification is arbitrary and therefore the exact proportions not significant. The important outcome is that a good share of this generation is essentially clean-legged like the Tosa-fowl ancestor, and an approximately equal proportion is heavily booted like the Cochin ancestor, while the rest are feathered to an intermediate degree like the parents.

4. FOOT COLOR.—This has a curious way of changing during the early months of the individual. White is often represented in the young by yellow. A "slate blue" or "bluish black" occurs; this may be a form of the willow from which yellow has been extracted.

First hybrid generation. Of 5 individuals two are recorded as white, one as yellow, one as willow, and one as slate blue. Here is practically equal frequency of the light and dark types.

Second hybrid generation. Fifty-three individuals give the following distribution of foot color:

Color.	$f.$	Per cent.	
White..............	11	20.8	51
Yellow.............	16	30.2	
Willow.............	20	37.7	49
Slate or bluish......	6	11.3	
Total.........	53	100	

This shows a practical equivalence of light and dark foot colors as in the first generation. The interpretation of this result must be left for later studies.

5. CORRELATION OF CHARACTERISTICS.—Considering only the three characteristics of plumage color, booting, and foot color, and assuming that game color and boot are dominant and light and dark feet equally apt to occur, we find the following calculated and actual frequency of each combination (actual percentage is in *italics*):

```
Game plumage....... 75%  69.8% ┌ Booted............56.3%  58.5% ┌ Light feet..........28.1%  30.2%
                                │                                │ Dark feet...........28.1%  23.5%
                                │ Non-booted........18.7%  11.4% ┌ Light feet..........9.4%  5.7%
                                │                                │ Dark feet...........9.4%  5.7%
White plumage......25%  30.2% ┌ Booted............18.7%  28.5% ┌ Light feet..........9.4%  13.2%
                                │                                │ Dark feet...........9.4%  15.1%
                                │ Non-booted......... 6.2%  1.9% ┌ Light feet..........3.1%  0.0%
                                │                                │ Dark feet........... 3.1%  1.9%
```

Considering the intrinsic difficulties of classification due to the partial blending of characteristics, there is a fairly close correspondence between the calculated and the actual. This result proves that there is little if any necessary correlation between the characteristics in question; they may combine in a chance fashion in the second hybrid generation.

CONCLUSIONS.

The inheritance of color in this cross between a white and a game-colored breed is remarkable in that white is not dominant—as is usually the case—nor recessive; but inheritance is particulate in the heterozygote, producing barred offspring. Segregation nevertheless occurs in the second hybrid generation, but the extracted whites and game colored birds are, for the most part, no longer as pure in color as their grandparents were. The germ cells are no longer perfectly pure—they have become infected by contact with the opposite quality.

The long-tailed characteristic behaves in inheritance like a unit character—in no wise different from plumage color. One can not help doubting whether it originated by any different method from that in which the diverse colors of poultry have arisen.

Foot feathering is dominant here as in many other cases; yet the dominance is incomplete. The germ cells of the second hybrid generation are no longer pure.

The White Cochin has no sexual dimorphism in plumage color, while the Tosa fowl is strongly dimorphic. Every one of the first hybrids is dimorphic in plumage coloration, the two sexes resembling, except for the white, respectively the female and the male Tosa fowl. It is striking to see how from a germ cell of the male Tosa fowl either a bird colored like a male Tosa or a bird colored like a female Tosa may arise. The male germ cells contain the Anlagen not only of the male characteristic but also of the female characteristic (Darwin, 1876, Chapter XIII).

Series X.—Dark Brahma and Tosa Fowl.

STATEMENT OF PROBLEM.

This series was undertaken primarily to test inheritance of secondary sexual characteristics and the possibility of transferring them from one sex to another.

THE RACES AS A WHOLE.

The Dark Brahma male and female have been described at page 32; the Tosa fowl, male and female, at pages 43, 44. Each race has a strongly marked sexual dimorphism in plumage color. The males have feathers of a more uniform color; the female Dark Brahma has penciled feathers; the female Tosa fowl has mossy feathers with prominent light shafting.

TABLE OF CHARACTERISTICS.

No.	Characteristic.	Tosa fowl. Female.	Tosa fowl. Male.	See page—	Dark Brahma. Female.	Dark Brahma. Male.	See page—
1	Shafting	Present	Absent	..	Absent	Absent	..
2	Lacing on hackle	Present	Present	..	Present	Present	..
3	Lacing elsewhere	Absent	Absent	..	Absent	Present	..
4	Penciling	Absent	Absent	..	Present	Absent	..
5	Red wing-bar	Absent	Present	..	Absent	Present	..
6	White wing bow	Absent	Absent	..	Absent	Present	..
7	Comb		Single	32		Pea	32
8	Earlobe		White, red edge	33		Red or bay	33
9	Iris color		Red	33		Yellow	33
10	Foot color		Willow	48		Yellow	33
11	Vulture hock		Absent	48		Present	34
12	Foot feathering		Absent	..		Present	34
13	Tail feathers		Long	44–48		Short	..

REMARKS ON THE CHARACTERISTICS.

1. SHAFTING.—In plumage, shafting is a light streak on the shaft and adjacent parts of the vane. Of the two parental races it occurs only in the female Tosa fowl (fig. 30). The light shaft-stripe is, however, widespread among females of certain dark or silvered races—*e. g.*, Silver Wyandottes, Silver-gray and Dark Dorkings, Silver Duckwing Games, and Silver Penciled Hamburghs. It crops out in many individuals where its occurrence is regarded by the "fancy" as a "defect." It is an original characteristic of poultry introduced from *Gallus bankiva*, whose female exhibits it conspicuously (fig. 39).

2. HACKLE LACING.—Among most broken-colored poultry the hackle feathers are unlike those of the rest of the plumage. Usually the hackle has a dark band in the center and is margined or laced by white—more rarely by yellow or red. In the female Jungle fowl (fig. 39) the feathers of nape and hackle have a black center (with broad, straw-colored shafting) and are laced with straw color. The male Jungle fowl has hackle feathers of a solid red color. In the descent of the domestic poultry, hackle lacing seems to have been transferred to the male sex also.

3. BODY LACING.—Few races of poultry exhibit lacing elsewhere than on the hackles. It is very prominent on the Indian female, but is not found on the Jungle fowl of either sex. In the Dark Brahma male (fig. 19) it occurs only on the saddle feathers. Whether its laced saddle is derived from the Indian or is due to a spreading, through correlation, from the hackles can not be said. Lacing is found on the breast of Game fowl and over much of the body of the female Dark Dorking. Among certain derived races, such as the Spangled Polish and the Laced Wyandottes, it affects nearly the whole plumage and is very conspicuous.

4. PENCILING.—This may be defined as a concentric repetition on the feather of alternating bands of the lacing and the ground color. In the hackle of the female Jungle fowl the straw color of the lacing is repeated in the center, the two light areas being separated by a black band. In the female Indian fowl the feathers of the throat are laced, but lower down on the larger back-feathers and on the wing bows, there is a second or inner lacing—i. e., the wing is penciled;* consequently penciling may be said to be a fundamental form of coloration in the genus *Gallus*. Penciling occurs widespread among the derived or secondary races of poultry, particularly in the "partridge" varieties. A curious modification of penciling is the straight transverse barring of the feather familiar in the Barred Plymouth Rock and Penciled Hamburghs.

5. RED WING-BAR.—The wing-bar is formed by the lower wing coverts, usually the first to third rows above the remiges or flight feathers. In the male of many races of fowl these differ from the more proximal rows. In the Dark Brahma male they have white and red in addition to black. The wing-bar has probably been derived by the Dark Brahma male from the Indian fowl. In the male Tosa fowl the lower wing coverts are tipped with red, but they show no white.

6. WHITE WING-BOWS.—The wing-bow is formed by the upper or proximal rows of wing coverts—i. e., above the third. These coverts are frequently of a different color from the wing bar. They are red in the male Indian and Malay, but they are white † in the male Aseel.‡ The white wing-bow of the Dark Brahma has probably been derived from this source. The wing-bow of the male Tosa fowl, like that of the Jungle fowl and Games, is red.

8. WHITE EARLOBE is a derived color, the primitive condition being red (page 33).

9. IRIS COLOR.—The origin of the yellow eye of the Brahma has been discussed at page 33. The red eye of the Tosa fowl is found in most Games and is the prevailing color among domestic poultry.

MATERIAL.

Mother.—No. 121, Dark Brahma Bantam (fig. 18).§ She is a beautifully penciled bird, with horn-colored beak, pearl-colored iris, prominent vulture hocks, and booted down to the outer two toes. To test her purity, she was bred for a month to No. 122, Dark Brahma male, also from Mr. Hodges. Their offspring died before hatching except one (No. 146 ♂), which is a typical Dark Brahma.

---

* Compare Wright, 1902, p. 334, and American Standard of Perfection, 1905, p. 207, figure.

† According to Ludlow's painting in Wright, 1902, opposite p. 326.

‡ Since the above was written I have purchased a male Aseel which has dark coverts tipped with white.

§ Weight 1,300 grams, received February, 1905, from Mr. F. H. Hodges Red Bank, New Jersey, marked F. H. H., No. 66, also No. 338.

Father.—A Tosa fowl bred at the station, No. 8A, "General Oyama," referred to at page 46.

RESULTS.

The produce was 5 females and 16 males (fig. 40). They are all blocky birds, very different from the Tosa fowl, but longer than the Brahma. The maternal or Brahma type is, however, predominant. Only the first generation of hybrids has been reared.

1. SHAFTING.—The male hybrids are mostly without shafting on the feathers of the back and the wing coverts. Two, however, show clear yellow shafting on these feathers, and in two others the shafting is a light buff color. The female hybrids have these feathers shafted. Shafting is dominant in the female hybrids. It is doubtfully transferred to some males.

2. HACKLE LACING.—This showed on all hybrids of both sexes.

3. BODY LACING.—In the male hybrids the saddle feathers and sometimes the tail coverts are laced with yellow as in the Brahma. Such lacing does not appear on the female. Lacing in the male sex appears to be dominant.

4. PENCILING.—This appears as typical penciling or as barring on the back and saddle and on the exposed parts of the secondaries of the female hybrids. It does not appear on the males. Penciling seems to be dominant over mossiness and to be confined to the female sex.

5. RED WING-BAR.—This is present in all of the first hybrid males, but the red is deeper and spreads farther over each feather than in the Dark Brahma, the red of the Tosa fowl having its effect. The female is without wing-bar as in the female parents.

6. WHITE WING-BOW.—Of 13 hybrid males four show no white in the upper wing coverts (fig. 40); but one of these has a light buff bow—a tendency toward white. The others have a small amount of white, which is derived from the Dark Brahma. The white has, however, been clearly reduced in amount. The interpretation of this result must await further breeding.

7. COMB.—In every hybrid the comb is pea, proving the dominance of that form over the single. The pea is, however, often atypical, the lateral ridges being rudimentary. Dominance is not always perfect.

8. EARLOBE COLOR.—Every hybrid shows some white, as in the Tosa fowl; but this white tends toward yellow—a much diluted red. White seems to dominate, but, if so, the dominance is imperfect.

9. IRIS COLOR.—This is red in the hybrids; but in two cases the red approaches orange. The iris color of the Tosa fowl is dominant, but imperfectly so.

10. FOOT COLOR.—Of 21 hybrids, all males (16) show yellow feet and all females (5) willow feet. This dimorphism is not found in the parent races.

11. VULTURE HOCK.—The hybrids show a tendency toward long feathers hanging over the heel (fig. 40). In one case these had reached a length of 105 mm. by six months; in another, about 90 mm. In other cases these

feathers are much reduced from the Brahma type, and in one or two cases it is doubtful if they are present. We have to do here either with a blending characteristic or else a very imperfect dominance of the vulture hock.

12. FOOT FEATHERING.—This is always present in the hybrids, but is usually less heavy than in the Dark Brahma (fig. 40). Booting is dominant, but is imperfectly so.

13. TAIL FEATHERS.—As none of the hybrids are over six months old, it is impossible to report fully on the inheritance of this characteristic. While in some male hybrids the tail feathers already surpass in length the middle tail feathers of the adult Brahma parent and are still growing, in no case have they made the extraordinary growth of the Tosa fowl.

CONCLUSIONS.

METHOD OF INHERITANCE.—The color characteristic of shafting and penciling in the female, and body lacing, red wing bar, and white wing bow in the male, appear to dominate in the respective sexes; but dominance, if such it is, is always imperfect, in that traces of the opposite allelomorph are sometimes found. Furthermore:

Red eye color dominates over yellow (not always perfectly).

Booting is dominant over clean leg.

Earlobe color is something of a mixture.

Vulture hock is sometimes very imperfectly "dominant."

The length of tail feathers is perhaps a blend.

SEX IN INHERITANCE.—For the most part a sexually dimorphic characteristic is inherited only by the proper sex. In the hybrids of this series, however, shafting seems to have been partially transferred from the female to some males. Most peculiar is the inheritance of foot color, where all the female hybrids show the willow foot of their father, and all male hybrids the yellow foot of their mother.

Series XI.—Frizzle and Silky.

STATEMENT OF PROBLEM.

This series of crossings was made to learn the inheritance of the allelomorphs given below.

THE RACES AS A WHOLE.

The origin of the Frizzle fowl (figs. 41 and 42) is not definitely known. Darwin (1876, Chapter VII) states that they are not uncommon in India, and Temminck states that they are domesticated also in Java, Sumatra, and all the Philippine Islands, being prevailingly white. They must have been brought to Europe early, since they are described by Aldrovandus in 1645 from a specimen sent him from Parma. Willoughby, in his Ornithology (1676), says that he had seen them in England. The recurving of feathers is found in many species of birds. It usually occurs on the neck, where it forms a ruff; more rarely over the entire body. Frizzled canary birds are occasionally exhibited. Frizzling is probably morphologically related to "rough coat" in mammals. The frizzled characteristic is a typical sport.

The Silky fowl (fig. 43) is likewise of great antiquity. Marco Polo saw it in Asia in the thirteenth century (teste, Dürigen, 1886, p. 298). Gesner described it in 1555. It is a native of eastern India, coming, according to Blyth (Tegetmeier, 1867, p. 221), from China, Malacca, and Singapore. A condition allied to silkiness (described below at page 57) is found in other races of poultry, particularly, as the following statements show, in the Cochins.

Tegetmeier (1867, p. 46) says:

The singular variety known as Silky Cochins, or sometimes as Emu fowls, is simply an accidental variation of plumage which occasionally occurs and which may be perpetuated by careful breeding. The cause of the coarse fluffy appearance of these remarkable fowls is to be discovered in the fact that the barbs of the feathers instead of being held together by a series of hooked barbules (so as to constitute a plane surface, as occurs in all ordinary feathers) are perfectly distinct, and this occasions the loose fibrous silky appearance from which the fowl obtains its name.

An engraving of such a feather is given by that author at page 224.

Wright (1902, p. 255) states that he has seen no Emu fowl "now for twenty years," and makes the suggestion that this entire "silkiness" of feather is the extreme limit, perhaps, of the kind of plumage which gives fluffiness to the leg region of American Buff Cochins.

The fluff of Cochins and Brahmas has indeed many points of similarity in structure with the feathers of the Silky. In one feather from the abdomen of a Brahma hen, whose shaft is 35 mm. long, I find the barbs very long (up to 30 mm.) and not connected together. Each barb bears, proximally, two rows of short, flat, hook-shaped barbules alike on the two sides. Beyond, there are a few short barbules that taper to a hair-like apex. Still more distally on the barb the barbules may attain a length of 5 mm., be altogether devoid of hooklets, but show a segmented condition as in the Silky. Far from my preconceived notion, I find few intergrades between the short barbules and the others. The more proximal of the long barbules are the longest of all, and the short barbules (which rarely exceed 0.5 mm. in length) also occur here scattered among the long ones. There thus seems to be a discontinuity between the two kinds of barbules, and this harmonizes with the view that the long barbule is a mutational form of the more typical short barbule.

As to the relation of the plumage of the Silky fowl to the fluff of Cochins, I have formulated the following hypothesis: Long and short barbules are two dimorphic forms found among birds. This dimorphism has been recognized in the terminology "down feathers" and "contour" + "quill" feathers. Down feathers may or may not have a shaft; they have barbs, and usually barbules, the latter being long and devoid of cilia or hooklets. In the contour and quill feathers of most birds the short barbules alone are present. But in some birds the barbules are long and devoid of cilia or hooklets as in the Ratitæ (ostrich, emu, cassowary, etc.). In poultry the down feathers are characterized by absence of hooklets, and the ventral abdominal feathers of poultry belong to this category. In the Silky fowl the contour feathers, in the strict sense, are absent, or rather they have gained long hookless barbs, and consequently have become in so far down feathers. But the feathers of the Silky fowl have one new characteristic not found in any other long-barbed forms, namely, the bifurcation and anastomosis of the barbs (page 58).

TABLE OF CHARACTERISTICS.

No.	Characteristic.	Frizzle fowl (Game).	Silky fowl.
1	Plumage color	Dark, black, red, and buff.	White.
2	Comb form	Rose.	Single.
3	Shaft of contour feather.	Recurved.	Straight.
4	Barb length.	Short.	Long.
5	Barb form	Twisted about long axis.	Straight.
6	Number of toes.	Four.	Five or six.
7	Skin color.	White	Black.
8	Crest	Absent	Present.

REMARKS ON THE CHARACTERISTICS.

1. PLUMAGE COLOR.—This characteristic is very variable in Frizzles, owing to the fact that fanciers have established no color "varieties," although an effort is now being made in that direction.* As stated below in detail, my Frizzles were of varied and mixed colors.

2. COMB FORMS.—The "American Standard" calls for single comb in the Frizzle and rose, or rather strawberry, comb in the Silkies. My Frizzles have, on the contrary, a rose comb, and my Silkies either a single comb or a rose comb, the Silky being impure in respect to this characteristic.

3–5. FEATHER FORM.—In the Frizzles the contour feathers have the shaft curved so that its outer surface becomes concave. This is most striking in the neck region, where a ruff is formed (fig. 41). The wing primaries are modified in another direction, since in them the barbs, in groups of 4 to 8, are twisted in corkscrew fashion about their own axis and through 180° or more; consequently the gray surface, which is normally next the body, comes to lie outermost. Such a twisting of the barbs sometimes occurs in primaries of non-frizzled races; particulary I have found it in the eighth primary of some Houdans. The barbs of the remiges of the Frizzles are mostly short, and in some cases are lacking altogether, being easily broken off.

The feathers of the Silky fowl are remarkable in all parts of the plumage. The contour feathers, as already stated, are down feathers, whose shaft is

---

*The new "American Standard of Perfection," published by the American Poultry Association, 1905, p. 248, directs that color should be "solid—black, white, red, and bay admissible, provided the birds match when shown in pairs, trios, and pens."

usually delicate but not otherwise atypical. The barbs are, on the other hand, remarkably long. Thus in a contour feather, from the middle of the dorsal region, whose shaft is 25 mm. long, the prevailing length of barb is 35 to 45 mm. The barbs are, moreover, remarkable in that they frequently bifurcate, even repeatedly. In a feather before me, one barb, taken at random, undergoes bifurcation four times. As the branches are not all in one plane, the feather becomes exceedingly fluffy. At the proximal end of the shaft the barbs arise parallel and produce an imperfect web close to the shaft, but marginally the web is lost. Distally on the shaft the barbs arise more irregularly from the shaft, often bifurcating almost immediately, so that no web or vane is formed. The barbs may also anastomose.

The barbules are not less strikingly modified than the barbs. They attain a length of from 1 to 2 mm. Moreover, it is not possible here, as in other races, to distinguish between a distal series of barbules carrying a row of hooklets or cilia and a proximal series without hooklets but with a folded edge into which the hooklets of the distal barbules catch. This impossibility is due, first, to the fact that the barbules are not in *two* series merely, but may arise in three planes, or irregularly; also, morphologically, all the barbules on the barb are alike. They are all segmented like the ordinary distal barbule, and the hooklets are represented by minute thickenings at the end of each segment. As a consequence of this structure the barbs do not hang together to form a vane and the fluffiness is still further exaggerated.

The quill feathers of the wing (remiges) and tail (rectrices) of the Silky are modified, but to a less degree. Primaries, secondaries, and coverts are all affected. The proximal part of the vane is nearly normal; the distal part has barbs of twice to thrice the normal length. The barbs may bifurcate repeatedly and even anastomose in the plane of the vane. The barbules also are modified, being much shortened. Proximal as well as distal barbules may carry hooklets, as is seen in the middle part of the feather. In the proximal part of the feather, on the other hand, the proximal barbules are without hooklets. The feathers of the tail have the web even more broken up than those of the wing.

The silky condition of the feather is a characteristic that is either entirely new (progressive in de Vries's sense) or possibly latent (in de Vries's sense) in typical fowl, so that its appearance in the Silky is a case of "degression" (de Vries). If the former, we should expect, according to de Vries, the offspring between a Silky and a non-Silky to show a mosaic of the parental feather characteristics and a non-Mendelian inheritance of silkiness; if the latter, a recessiveness of the varietal characteristic of silkiness and its Mendelian inheritance.*

---

* De Vries, 1905, p. 280: "The character of the species is dominant in the hybrid, while that of the variety is recessive." On the latter of the two assumptions made above, plain plumage is the species character; silky plumage, the varietal.

6. NUMBER OF TOES.—This is constantly four in pure-bred Frizzles. In Silkies a fifth toe is always present. The extra toe frequently has a double nail, or the division may be complete, resulting in six toes.

7. SKIN COLOR.—In the case of the Frizzle the skin is white, sometimes tinged with yellow pigment. The skin of the Silky is notoriously blue-black. This is a clear case of melanism, and since early times has been associated with the other peculiarities of the Silky. The melanic condition affects the periosteum also. It is remarkable that despite this excess of pigment rendering black the internal tissues, skin, leg scutes, comb, and wattles, the plumage should be always *white*.

MATERIAL.

Mothers.—Four Frizzles (Nos. 14A, 18A, 19A, and 20A), hatched May, 1904, from eggs obtained from Dr. A. G. Phelps, of Glen Falls, N. Y. All have rose combs and slightly booted feet. No. 18A is peculiar in that the feathers on head and neck are sparse and small (fig. 42). In general color the hens vary; 14A is prevailingly dark brown; 18A is yellowish; 19A is light brown, and 20A is mixed black, yellow, and red. A male Frizzle from the same lot of eggs was highly colored red and black.

Father.—A white Silky cock (No. 24A, fig. 43), likewise hatched from eggs sent in May, 1904, by Dr. Phelps.

The Silky cock and Frizzle hens were mated from January 16 to April 14, 1905. Trap nests were not used, so that I could distinguish mothers only by the form of the eggs. The egg of 18A was very peculiar and was early identified. A certain proportion of the offspring can not be assigned to any particular mother.

RESULTS.

Only the first hybrid generation has been obtained.

1. PLUMAGE COLOR.—Of 32 hybrids, 7 (22.6 per cent) are white (showing some buff in six cases) and 25 (77.4 per cent) are dark. No. 18A apparently produced only dark birds, largely dead-black. The others produced in part white hybrids (fig. 44), but mostly pigmented ones. The result is not what we should have expected. If white were recessive, 0 to 50 per cent, if dominant, 100 per cent, of the offspring should be white. Moreover, the Silky is doubtless homozygous in respect to color, since (1) Silky fowls are carefully bred for white color, and (2), bred to a hen of its own strain, it has produced only white birds. I conclude, therefore, that the white plumage color is not always dominant over the black, red, and yellow of the Frizzle. The matter will be further investigated.

2. COMB.—In all cases the rose comb of the Frizzle dominated over the single comb of the Silky (fig. 44).

3–5. CURVING OF SHAFT, BARB LENGTH, AND BARB FORM.—These are all correlated in the first generation. Of 10 mature birds, 6 are typically frizzled and 4 have flat feathers. Assuming frizzling to be dominant, non-frizzling recessive, and that all my Frizzle fowls are heterozygous, we should expect 50 per cent frizzled offspring. The result accords well with these hypotheses. None of the hybrids show any trace of silkiness. Silkiness is recessive as against non-silkiness.

This result is striking and has been observed by others. Tegetmeier (1867, p. 224) bred Silkies to other varieties and found that "the chickens produced seldom had the silky feathers, but were clothed in plumage of the ordinary character." Mating these hybrids together he got among plain feathered offspring "one covered with feathers like those of the Silk fowl," but with black plumage. Darwin (1876, Chapter VII) had previously bred a white Silk hen to a Spanish cock; "none inherited the so-called silky feathers."

6. NUMBER OF TOES.—Thirty hybrids gave the following distribution of characteristics:

Characteristic.	$f.$	Per cent.
4 toes, both feet..............	7	23.3
4 and 5 toes.....	9	30.0
5 toes, both feet................	14	46.7
Total......................	30	100.0

Here, as elsewhere in this paper, the inheritance of extra toe is difficult to account for on the Mendelian principle of dominance.

7. SKIN COLOR.—All hybrids have a black skin. Tegetmeier (1867, p. 224) got the same result.

8. CREST.—So far as noted, all mature hybrids have a well-marked crest, but it is somewhat smaller than that of the Silky.

CONCLUSIONS.

A final conclusion as to dominance must await the production of the second generation of hybrids. The following (in *italics*) appear to show Mendelian dominance over the corresponding allelomorphs:

Rose comb   vs. Single comb.
Frizzle feathers vs. Plain feathers.
Black skin   vs. White skin.
Crest   vs. Plain head.

Plumage color and number of toes are unit characters, but behave peculiarly. The dominance of the crest is imperfect.

Series XII.—Single-comb White Leghorn Bantam and Black-breasted Red Rumpless Game.

STATEMENT OF PROBLEM.

This cross was undertaken primarily to test the inheritance of rumplessness, and secondarily of the more primitive game coloration against white plumage color.

THE RACES AS A WHOLE.

The White Leghorns have been described at pages 37 and 39. The Black-breasted Red Game closely resembles the wild Jungle fowl in color (figs. 45 and 46).

TABLE OF CHARACTERISTICS.

No.	Characteristic.	Single-combed White Leghorn.	Discussed at page—	Black-breasted Red Rumpless Game.	Discussed at page:
1	General color....	White..........	18	Black and red.....	..
2	Beak color.......	Yellow..........	19	Black............	..
3	Uropygium......	Present.........	..	Absent...........	..
4	Foot color.......	Yellow..........	20	Willow...........	48

REMARKS ON THE CHARACTERISTICS.

UROPYGIUM.—The absence of uropygium is a characteristic that has long been known among fowl, but there seems to be little knowledge of its morphology. In ordinary fowl there are five free caudal vertebræ, followed by a fused portion—the uropygial bone. In the case of a rumpless Game female (No. 119, fig. 45) dissected by me, there are two unsymmetrically formed and intimately fused vertebræ behind the fifteenth synsacral—the posterior limit of the sacral vertebræ. That there are two is shown by distinct transverse processes with spaces of the passage of the nerves. Behind these is a knob of bone about 1 mm. in diameter. These three elements constitute the entire caudal skeleton. It is profoundly reduced from the normal.

Rumplessness may be found in any race. It has cropped out in two of the 800 fowl bred at this station in the past year—hybrids derived from the Minorca-Polish and the Leghorn-Houdan crosses. It seems like a misuse of the term *breed* to speak of a "Rumpless breed," as poultry books do.

The characteristic is referred to by Aldrovandus in 1645, by Temminck, and by other early writers. Its origin has been ascribed to Persia, to Ceylon, and to China; doubtless it occurs in all these places as well as in many others. Taillessness early appeared among fowls in America. Clayton (1693, p. 992) asserted that he had observed that in "Virginia" most of the cocks and hens were without tails, and Wright states that he was informed by a West Indian in 1872 "that the greater number of fowls in his own neighborhood had no tails." Darwin (1876, Chapter VII) refers to this characteristic and states that one bird he examined had no oil gland; the same is true of the three rumpless Games that I have had. Among the poultry books that describe the "breed" quite fully are Tegetmeier (1867, pp. 230–232), Baldamus (1896, pp. 170–172, "Kaul oder Klütthühner"), Dürigen (1886, pp. 98–100), Wright (1902, p. 481), and Weir-Johnson-Brown (1905, pp. 1016–1017).

Regarding the inheritance of this characteristic, statements are not in accord. Tegetmeier (p. 231) says:

> A friend of mine purchased a successful pen [of Rumpless fowl] at a poultry show, taking them away to a walk where no other fowls ever trespassed, and yet the chickens were, in a considerable number of instances, furnished with fully developed tail feathers, being not rumpless. On inquiry of the previous owner, he stated: "Mine have always done so from the first time I kept them; but the tailed birds will very probably produce rumpless chickens." Three such birds were purposely retained, and they produced the next year more than twenty youngsters, all of which but one were rumpless and destitute of tail feathers.

The foregoing experiment would seem to prove that the rumpless parents were heterogametous, and that while rumplessness is dominant the recessive condition of tail is here prepotent (Castle, 1905). Darwin (1876, Chapter VII) possessed a rumpless bird which "came from a family where, as I was told, the breed had kept true for twenty years; but" he adds, "rumpless fowls often produce chickens with tails." The breeding true of a character may mean either that it is dominant and homogametous in this respect or that it is recessive. Dürigen (1886, p. 99) states that a rumpless cock mated with a tailed hen produces not exclusively rumpless, but a fair percentage of them, and Wright (1902, p. 481) says that "a Rumpless fowl crossed with any other generally produces a large majority of Rumpless birds." All of the foregoing results are consonant with the conclusion that rumplessness is typically dominant, but that the recessive full tail may be prepotent.

MATERIAL.

The *mother* was the White Leghorn bantam No. 127 discussed at page 39. She is heterozygous and contains black gametes.

The *father* (No. 117, fig. 46) was one of three rumpless bantams obtained from Dr. A. H. Phelps, of Glen Falls, New York. Two of these were typical Black-breasted Red Games; they lack oil glands and weigh about 1,000 grams each.

RESULTS.

Only the first generation of hybrids has been so far obtained.

GENERAL PLUMAGE COLOR.—Of 24 hybrids 12 were white or prevailingly so (fig. 47). Usually, however, more or less black and more rarely some buff was present. The other 12 were either black-and-white barred (and these were all males) or black with more or less reddish. As we have seen, the white mother contains recessive black or black-and-white, so that the result accords with the expectation of only 50 per cent white.

BEAK COLOR.—In the hybrids the beak is sometimes yellow, sometimes black, sometimes black-and-yellow streaked.

UROPYGIUM.—Of 24 hybrids the uropygium is normal in 23 (fig. 47). One chick taken from the egg is recorded as without tail, though tail gland is present. It is doubtful if much stress may be laid on this record, as the uropygium is always very small in the unhatched bird. We may exclude it from present consideration. This whole result was unexpected because opposed to the earlier observations. It leads to the provisional hypothesis that rumplessness is recessive in my strain. If full tail is recessive, then in my strain the recessive condition is prepotent. Further discussion must be deferred until the second hybrids have been bred.

FOOT COLOR.—This was yellow in about half of the cases and willow or dark in the other half. Recalling that the White Leghorn is heterozygous, the result favors the hypothesis that yellow is dominant over willow.

CONCLUSIONS.

White plumage color seems to be dominant over game color. The hypotheses seem to be warranted that yellow beak and foot color are dominant, and that rumplessness is recessive in this strain.

Series XIII.—Black Cochin Bantam and Black Breasted Red Rumpless Game.

STATEMENT OF PROBLEM.

This cross was primarily to test the inheritance of rumplessness, and secondarily of black against red plumage color.

THE RACES AS A WHOLE.

Concerning the Rumpless Game see page 61. The Black Cochins are discussed at page 39.

TABLE OF CHARACTERISTICS.

No.	Characteristic.	Black Cochin Bantam.	Discussed at page—	Black-breasted Red Rumpless Game.	Discussed at page—
1	General color....	Black.........	39	Red with some black.	62
2	Uropygium......	Present.......	63	Absent.............	63
3	Iris color........	Dark brown ..	..	Red streaked with yellow......	..
4	Vulture hock ...	Present.......	39	Absent..	..
5	Foot feathering..	Present.......	34	Absent.............	..

MATERIAL.

Mothers.—The Black Cochin Bantams Nos. 129 (fig. 26), 130, 131, and 132 were the same as those referred to at page 39.

Father.—The Rumpless Game is No. 117, referred to at page 62 (fig. 46).

RESULTS.

Only the first hybrid generation has been produced.

GENERAL PLUMAGE COLOR.—Of 24 hybrids all were prevailingly black. Among 18 of those that hatched 8 showed some red. This red is chiefly found as a lacing on the hackle feathers or a peppering on the wing coverts,* throat,† and outer margins of the remiges.‡ This seems to point to the hypotheses that while black dominates over red the dominance is sometimes imperfect. When red occurs it occurs on those feathers that normally contain red in the Game, and on that part of the feather that is red in the Game.

UROPYGIUM.—This is invariably present, apparently fully developed.

IRIS COLOR.—All the hybrids have dark-brown eyes; only one shows a trace of red. The hypothesis seems justified that in this case dark-brown iris pigment is dominant over red and yellow.

VULTURE HOCK.—This is always absent. In only a single case§ are the feathers slightly elongated on the hock.

FOOT FEATHERING.—Every chick that hatched has the foot and at least one toe booted. In some cases this booting is much reduced as compared with the Cochin parent. Booting is dominant, but not always completely so.

CONCLUSIONS.

In this cross of black *vs.* red, black appears to be dominant, although imperfectly so. The two colors do not blend, however, but red appears in a particulate fashion, usually in the parts of the plumage that have normally least black pigment. It is as if there were a struggle between the two pigments and red overcame black where black was weakest.

The presence of tail in the first hybrid generation is confirmatory of the results of the preceding series. Rumplessness is apparently recessive.

Brown iris color appears to dominate over the older red, and booting dominates over the ancestral clean-footed condition.

---

* Nos. 589 ♂ and 798 ♂.   † No. 587 ♀.   ‡ No. 577 ♀.   § No. 651 ♂.

D. GENERAL DISCUSSION.

INHERITANCE OF PARTICULAR CHARACTERISTICS.

COMB FORM.

The comb is a characteristic that has had its origin in the genus *Gallus*. It consists of a mass of uncovered erectile tissue—a tissue present in many species of birds. The primitive form of the comb is the single comb seen in the wild species of the genus *Gallus*, and in most domestic races. This may be modified in two directions : First, in the direction of lateral repetition of the comb giving rise to the pea comb,* and, in an extreme case, to the rose comb (of which the walnut comb of the Malays is a special modification); second, in the direction of reduction of the modified comb producing the races with mere papillæ (Houdan, Polish, La Flèche, etc.) or that are entirely combless (Breda fowl). That the rose comb is a modification of the same sort as the pea comb but carried to a greater extreme is indicated by the fact that the rose comb often shows five parallel ridges (instead of the more usual irregularly scattered papillæ) and that in the female the rose comb sometimes consists of three ridges as in the male pea comb.

When single comb (Minorca, fig. 4) and pea comb (Brahma, fig. 19) are crossed, pea comb is dominant (p. 35). The median ridge is, however, in the hybrid high for a pea comb and the lateral ridges are usually reduced (figs. 20, 21). When single comb (Leghorn) and rose comb (Minorca) are crossed, rose comb is dominant (p. 30). When single comb (Minorca or Leghorn) is crossed with the paired rudiments of a comb found in the Polish and Houdan fowl, a Y-shaped comb results (pages 10, 22, 28, fig. 8). This Y comb is of great interest. It was obtained by Bateson and Punnett (1905, pp. 108, 112–114) in some of the offspring of (single-comb Leghorn × rose-comb Dorking), crossed with (single-comb Leghorn × walnut-comb Indian); and also in one of the offspring of a single-comb Leghorn crossed with [(single-comb Leghorn × walnut-comb Indian) × (single-comb Leghorn × rose-comb Dorking)]. In Bateson and Punnett's cases the splitting was evidently nearly complete, forming an O-shaped comb, or the "cup comb" of Darwin (1876, Chapter VII). The Y comb was obtained also by Hurst (1905, pp. 133, 135, 138, 140, 146). This was a single split comb when Leghorn and Houdan were crossed, and a rose split comb when rose-comb Hamburgh and Houdan were mated.

The interrelation of the different forms of comb—single, pea, walnut, Y, and V may, I think, be expressed in the following hypothesis : The pea comb and the walnut comb are composed of two elements—a median single comb and a pair of lateral combs. This hypothesis is supported by the

---

*The pea comb was doubtless a characteristic of the unknown feral ancestor of the Aseel-Indian group. But as the single comb is the dominant type in the known wild Jungle fowls the pea comb probably evolved from it.

following evidence. First, of teratology. Extraneous paired papillæ occasionally occur on the sides of the single comb in pure-bred races. These are known as "side springs," and are considered by fanciers as grave "defects." Now such side springs are morphologically equivalent to the lateral ridges of the pea comb. Second, there is the evidence of hybrid forms. Bateson and Punnett (1905 a) show that when pea comb and rose comb are crossed the second hybrid generation (F_2) gives single comb, as well as pea and rose combs. This result may be interpreted as due to the fact that the gametes of a pea-combed bird have either a tendency toward side-springs (= pea comb) or they have no such tendency (= single comb); and the gametes of a rose-comb bird have a tendency to produce two pairs of side combs (= rose comb) or else they have no such tendency (= single comb). When two gametes without the side-comb tendency come together in F_2 a single comb is produced. The necessity of assuming absence and presence of lateral combs strengthens the view that the pea comb is made up of two elements—median and lateral. If median comb and side-springs are distinct elements, then they should be independently inheritable. This result is realized on the one hand in the single comb, and, I think, on the other hand, in the cup comb (fig. 6), which consists of two side-springs without median comb. It is realized also in the V comb of the Polish fowl, which is a cup comb of which the anterior portion is typically not developed.

That the V comb represents the posterior portion of a cup comb is supported by the fact that it is not uncommon to find not one pair of papillæ merely, but two, three, or four pairs of papillæ in Polish fowl and in second-generation hybrids. A row of three or four papillæ on each side of the head is a close approach to a typical cup comb.

The incompleteness of the cup comb where a V comb is produced may be due to various causes. In the Polish fowl the upturned nasal process and absence of a bony ridge over the nostrils appear to be the cause of the absence of a comb there, and we have seen (p. 17) that the only undissociable characteristics in the second-generation hybrids of Minorca and Polish are those of high nostril and rudimentary comb. The second cause restricting the development of the cup comb to its posterior limits is the presence of a median comb anteriorly; this is the case of the ordinary Y comb. The Y comb is found in hybrids between single and V comb; the anterior portion of the comb is not suppressed here, because the bony roof of the culmen is completely developed, and the very presence of a large median comb there prevents the development of the side-springs at the same niveau. In the development of the comb of the hybrid there is, as it were, a struggle between the two elements of median and lateral combs. The Y comb assumes a great variety of forms, running the entire gamut from a single comb on the one hand to (1) a cup comb or to (2) a pair of papillæ on the other. I have already (p. 10) referred to the variation of the length of the stem of the Y, series

(1), from 100 per cent to nearly zero. The second hybrids of Polish or Houdans crossed with single combs illustrate series (2). We begin with a single comb having its posterior one-sixth split; next comes a comb having its posterior one-sixth split and anterior five-sixths single, but greatly reduced in height (fig. 50); next the same with the anterior portion reduced to an irregular carunculated mass having a slight median elevation (fig. 49), and finally a pair of papillæ only (fig. 48). In this series we have a fading out of the median portion, *pari passu* with the enlargement of the nostril, but the persistence of the side combs unimpaired. The side combs have been unable to enter the territory from which the median comb has been driven, because that territory is likewise untenable for it. These two series sufficiently demonstrate that the V comb represents the posterior portion of the cup comb.

That the cup comb represents merely the greatly enlarged lateral combs or side-springs is proven by the occasional presence of both median and cup comb on the same individual. In some races, as in the English type of Houdans, the median comb typically appears lying between the pair of cup-like side-springs, resembling the trunk of a butterfly between its wings. Among the heterozygous combs of the second generation of Minorca × Houdan or Minorca × Polish hybrids, instructive examples of persistence of both single comb and side-springs are especially apt to occur. Figure 52 shows this condition; there is a median comb anteriorly and a nearly typical pea comb posteriorly, except that the lateral ridges are atypically high. Thus the Y comb becomes explained as due to the presence of both single and lateral combs.

The question now arises, Is it possible to explain on Mendelian principles the production of a Y comb when median comb and lateral comb are crossed? In accordance with such principles we should have to picture the gametes of the single-comb and V-comb parents as follows:

Single comb.	V-comb.
Median element.......	No median elements.
No lateral elements ..	Lateral elements.

The allelomorphs are then median and no median, no lateral and lateral, and the positive characteristics are dominant. In the second hybrid generation the two dominant characters should be combined in nine-sixteenths of all cases; the two recessive in one-sixteenth, and one dominant with one recessive in three-sixteenths + three-sixteenths of the cases.

Another hypothesis is possible. Granting that the Y comb is no neomorph, but the sum of single and lateral comb, then the Y comb may be a case of particulate inheritance, the median comb being produced on the anterior and the lateral on the posterior part of the frontal region. In cases of particulate hybrids bred *inter se*, the offspring exhibits one or the other of the parental conditions each in 25 per cent of the cases and the heterozygous condition in 50 per cent. To decide between these rival hypotheses we have to appeal to the statistics of occurrence of the different forms of comb. All cases (Series I and II) are combined in the following table, showing distribution in the second hybrid generation:

Comb characteristic.	Expected.		Actual.
	On hypothesis of dominance.	On hypothesis of particulate inheritance.	
	Per cent.	*Per cent.*	*Per cent.*
Single comb............	18.75	25	30.1
Y comb	56.00	50	44.9
Lateral comb...........	18.75 } 25.00	25	25.00
No comb...............	6.25	0	0?

The foregoing table reveals several things. First, the actual distribution of comb form in the second generation accords better with the hypothesis of particulate inheritance than that of dominance of both single and lateral comb. That there is an excess of single comb and deficiency of Y comb is partly accounted for by occasionally counting a potentially Y comb but actually single (or nearly single) comb as a true single. Secondly, the hypothesis of dominance demands the occurrence of a fourth form—presumably no comb—in 6¼ per cent of the cases. No combless fowl was raised to maturity, and the only possible cases were seen in still very young or unhatched chicks. Probably no true combless bird appeared. From both of these considerations I conclude, provisionally, in favor of the theory that the Y comb is reproduced from the median and the lateral by *particulate* inheritance.

NOSTRIL FORM.

The sum of results in Series I, II, and III (narrow × high nostril) gives:

Generation.	Narrow and intermediate.			High.		
	f.	Actual.	Expected.	*f.*	Actual.	Expected.
		Per cent.	*Per cent.*		*Per cent.*	*Per cent.*
F_1................	102	99.0	100	1	1.0	0
F_2................	99	73.9	75	35	26.1	25
F_1 × narrow......	33	51.5	50	31	48.5	50

A close agreement exists between the percentage obtained in each generation and the expectation on the Mendelian theory, assuming that narrow nostril is dominant. The statistics do not, however, tell the whole story. In 36 per cent of the cases in the F_1 generation the nostril was wider than in the "narrow" ancestor. Even in the F_2 generation nearly half of the "narrow and intermediate" were of the intermediate sort. This intermediate form is evidence that dominance is imperfect and segregation is incomplete.

CEREBRAL HERNIA.

Cerebral hernia is, as already pointed out, a typical monstrosity. The distribution of its occurrence in crossing is as follows:

Crosses.	F_1.		F_2.		$F_1 \times$ plain.	
	Plain.	Hernia.	Plain.	Hernia.	Plain.	Hernia.
Minorca × Polish	66	0	75	23	34	0
White Leghorn × Houdan	24	0	34	11	25	†0
Houdan × Minorca	16	*0	..	..	..	..
Total	106	0	109	34	59	0
Percentage	100	0	76.1	23.9	100	0

* Excluding one case of egg embryo with cerebral vesicle.
† Excluding one egg embryo recorded as doubtful.

Cerebral hernia is inherited in Mendelian fashion with plain head dominant. Nevertheless, many of the plain-headed hybrids have the frontal eminence abnormally high—dominance is imperfect.

CREST.

The crest is independent of the cerebral hernia (pages 16–18). It is a widespread characteristic among birds, so common that it is not readily thought of as pathological but usually as ornamental. The distribution of its occurrence in crossing is as follows:

Crosses.	F_1.		F_2.		$F_1 \times$ plain.	
	Plain.	Crested.	Plain.	Crested.	Plain.	Crested.
Minorca × Polish	0	70	11	41	6	6
White Leghorn × Houdan	0	25	6	13	6	9
Houdan × Minorca	0	9	..	..	..	..
Frizzle × Silky	0	5	..	..	..	..
Total	0	109	17	54	12	15
Percentage	0	100	24	76	44.5	55.5

Crest is inherited in Mendelian proportions, and is dominant over crestless head. Even when the Silky is crossed with *Gallus bankiva* its crest is dominant (fig. 53). In this case the new characteristic, a positive variant, dominates over the ancient one; but the crest is diminished in the first generation; dominance is imperfect.

WHISKERS OR MUFF.

This is certainly a new character and a positive variant. The distribution of its occurrence in crossing is as follows:

Crosses.	F₁.		F₂.		F₁ × plain.	
	Absent.	Present.	Absent.	Present.	Absent.	Present.
Leghorn × Houdan	0	24	?	26	5	11
Houdan × Minorca	0	11	..	..	..	..
Total	0	35	?	26	5	11

Muffling is apparently dominant.

BEARD.

This is also a new, positive, variant. The distribution of its occurrence is as follows:

Crosses.	F₁.		F₂.		F₁ × plain.	
	Absent.	Present.	Absent.	Present.	Absent.	Present.
Leghorn × Houdan	0	23	?	12	3	8
Houdan × Minorca	0	10	..	..	..	..
Total	0	33	?	12	3	8

Beard is apparently dominant, but often imperfectly so.

FEATHER FORM.

Silkiness is a new characteristic and, approximating as it does the juvenile down condition, a negative one. When a Silky is crossed with a Jungle fowl the offspring are plain. Silkiness is recessive to non-silkiness—the retrograde to the progressive type.

Frizzling is likewise a new characteristic—a positive character added to the perfect feather. The distribution of the occurrence of silkiness and frizzling is as follows:

F₁.	No silkiness.	Silkiness.	Non-frizzling.	Frizzling.
Frizzle × Silky	10	0	4	6

The Frizzle fowl used were doubtless heterozygous. When non-frizzled birds are crossed *inter se* they produce only plain offspring. Frizzling is dominant over non-frizzling—the progressive over the primitive.

UROPYGIUM.

Rumplessness is a new characteristic and a typical negative variant. The distribution of its occurrence is as follows:

F_1.	Non-rumpless.	Rumpless.
Leghorn × Rumpless Game......	23	*0
Cochin × Rumpless Game.........	19	0
Frizzle × Rumpless Game.........	7	0
Nankin × Rumpless Game........	3	0
Total..................	52	0
Percentage.............	100	0

* One egg embryo doubtfully rumpless.

The new, *negative* characteristic is here completely recessive.

TAIL-LENGTH.

The long tail of the male Tosa fowl is a new, positive variant. The distribution of its occurrence in male hybrids is as follows:

Crosses.	F_1.		F_2.	
	Short.	Long.	Short.	Long.
Tosa × Cochin......................	0	3	?	?
Brahma × Tosa.....................	0	*16	..	..
Total........................	0	19	..	..

* The tails are, perhaps, more properly intermediate. While still growing at date of record, they grow slowly.

The new, positive characteristic is doubtfully dominant, possibly intermediate (*cf.* fig. 34).

VULTURE HOCK.

This bundle of strong feathers constitutes a new, positive characteristic. The distribution of its occurrence is as follows:

Crosses.	F_1.		F_2.	
	Absent.	Present.	Absent.	Present.
Minorca × Brahma...............	7	*0	..	..
Leghorn × Brahma...............	12	0	..	..
Black Cochin × Leghorn.........	12	(Small) 1	..	..
Leghorn × Buff Cochin..........	9	0	..	..
Tosa × White Cochin............	†3	†3	16	‡20
Brahma × Tosa...................	All grades.		..	..
Black Cochin × Rumpless Game.	§11	0	..	..

* One shows trace of enlargement of feathers. ‡ Seven recorded as slight.
† Females with vulture hock; males without it. § One case of trace of elongation of feathers.

The result is peculiar. Usually the vulture hock is absent in the first hybrids, indicating its recessiveness. In crosses with a particular race—Tosa fowl—however, there is no recessiveness. It is probable that the Tosa fowl is heterozygous in respect of this characteristic. The new characteristic is recessive, but imperfectly so.

FOOT FEATHERING.

Foot feathering, as the discussion on page 34 indicated, is a positive variation, new to *Gallus*, but not of a pathological sort. Common among wild, scratching birds, its occurrence in *Gallus* may be regarded as a case of degressive variation (de Vries). The distribution of its occurrence is as follows:

Crosses.	F_1.		F_2.	
	Non-booted.	Booted.	Non-booted.	Booted.
Minorca × Dark Brahma	1	40	..	..
Leghorn female × Dark Brahma male	4	15	..	..
Dark Brahma female × Leghorn male	0	25	..	..
Black Cochin × Leghorn	0	20	..	..
Leghorn × Buff Cochin	3	26	..	..
Tosa × White Cochin	0	7	7	48
Dark Brahma × Tosa	0	22	..	..
Frizzle × Silky	3	15	..	..
Black Cochin × Rumpless Game	0	21	..	..
Total	11	191	7	48

The foregoing statistics tell only a part of the story. Booting, when present, is frequently much reduced; one may regard absence of booting as the extreme condition. Booting is dominant, but usually imperfectly so.

EXTRA TOES.

The extra toe is a positive variation of a teratological sort. The distribution of its occurrence is as follows:

Crosses.	F_1.		F_2.		F_1 × normal.	
	No extra toe.	Extra toe.	No extra toe.	Extra toe.	No extra toe.	Extra toe.
Houdan × Leghorn	6	31	17	6	17	8
Houdan × Minorca	6	15	..	..	..	..
Frizzle × Silky	7	23	..	..	..	..
Total	19	69	17	6	17	8
Percentage	21.6	78.4	73.9	26.1	68.0	32.0

These results are peculiar. If both normal-toed and extra-toed ancestors were heterozygous in respect to toes, we should expect the result obtained in F_1. It is quite possible, though not probable, that this is true. Then extra toe would be dominant, although sometimes so imperfectly so as not to appear. In F_2 the parents were normal-toed, either because "normal" is recessive or because it is imperfectly dominant. All offspring should be normal-toed in the one case or give 100 per cent to 75 per cent extra-toed in the other. The result is not in accordance with either hypothesis. If there is any dominance in this generation it is of the *normal* toe. Bateson and Saunders (1902, p. 124), while concluding that extra toe is dominant, find "that the recessive foot character may sometimes dominate." Hurst (1905, p. 150) also got, in a cross between Leghorn and Houdan, some normal-toed offspring which, interbred, produced extra-toed progeny. He concludes that a usually dominant character may recede in certain individuals. There is danger here of straining Mendel's law. It is better to hold "explanations" in abeyance until the matter of inheritance of polydactylism has been more thoroughly investigated. Certainly the facts of inheritance of polydactylism in man can hardly be explained on Mendelian principles (Davenport, 1904). Polydactylism is at least not recessive. The new, positive, pathological characteristic holds its own against the older one.

SKIN COLOR.

The epidermis of poultry is everywhere covered by feathers except on the beak, face, and feet. The naked portions may, however, have a different color from the covered ones; consequently the correlation between general skin, beak, and foot color, although not absent, is not close. Thus, although the yellow beak and foot of the Leghorn are correlated with its yellow skin, the black legs and beak of the Black Minorca are not accompanied by a black skin. Not all exposed parts, even of the skin, are of one color, for the face, at least, may be red or white when the legs are black. Color of beak and foot are, on the other hand, closely correlated, individual variations of the one being usually associated with corresponding variations of the other. This correlation is doubtless the result of the similar cornification of the skin of beak and foot, whereas (excepting races with opaque white face) the vascular face and earlobes are white or red, according to a less or greater blood supply in them.

The pigmentation of the epidermis of poultry falls into three classes: (*a*) Without pigment or white; (*b*) yellow; (*c*) black. White skin is the commonest, even among poultry with black plumage and feet. Yellow skin is found in the Asiatics, derived from the Aseel-Malay ancestry, and is a characteristic of the White Leghorn. Black pigment occurs in the skin of the Silky fowl and the Negro fowl. Black pigment is to be regarded as a new variant and of the nature of a pathological sport—melanism. When black skin is crossed with white, black—the new, positive, pathological characteristic—is dominant (page 60).

MANDIBLE COLOR.

The prevailing types are black, willow, yellow, and white. Black is the primitive color on the Jungle side; yellow, on the Aseel group. When horn (Houdan) and yellow (Leghorn) mandible colors are crossed, the first generation shows the yellow of the Leghorn, which is dominant. When, however, the black beak of the Minorca was crossed with the yellow beak of the Brahma, the dark color dominated. The potency in the hybrid of beak color seems to follow this series: Black, yellow, horn. The most positive character, black, dominates all.

FOOT COLOR.

Four principal types are to be distinguished—willow, black, yellow, and white. Willow is primitive and white the most aberrant. The results are based on still insufficient data, but so far as they go they indicate that willow is dominated by yellow (p. 54), yellow by white (p. 24), and white by black (p. 28). The newer, negative characteristic, white, is dominant over the older yellow, but the new, positive characteristic of melanism dominates all.

IRIS COLOR.

Of the various forms, pearl to yellow is characteristic of the Aseel type; red, of the Game or *Gallus bankiva* type. Black has become associated with black plumage. The results, subject to revision, indicate that in poultry, as in man, iris color rarely blends, that red dominates pearl (page 38), and that dark brown dominates red. The new, positive variation of melanism seems to dominate all, although not always perfectly.

EARLOBE COLOR.

Red is primitive in both groups. White is a new variation, which is probably due to fat or other particles in the skin, and is consequently positive. Only in extreme cases is red wholly eliminated from the earlobe. In three series of crosses (V, VI, and X) of the red-lobed Dark Brahma and a white (and red) lobed race the earlobes were prevailingly red, but had some white at their centers. Likewise, in two series of crosses (VII and VIII) of the red-lobed Cochin and a white-lobed Leghorn, red dominated in the hybrids, but did not always perfectly exclude white. Red is apparently dominant, but very imperfectly so; some cases rather indicate particulate inheritance.

GENERAL PLUMAGE COLOR.

The original plumage as exhibited in the Jungle fowl is largely black and red; that of the Aseel type sometimes contains much white; but the pure white plumage must be regarded as a new negative variant. The outcome of crossing is complex.

WHITE VS. DARK—Three different results may be, under differing conditions, obtained.

Dominance of White.—This is the usual result. Two White Leghorns crossed by a black Minorca produced only white hybrids, but the female hybrids, at least, had some black feathers. White Leghorns crossed with Houdans gave only white. White Leghorns crossed with a Red-backed Game had white offspring with some buff on breast. On the other hand, the white color of the Silky dominates over the dark color of the Frizzle (Series XI) in about only 23 per cent of the hybrids. Bateson and Saunders (1902, pp. 108-109), dividing all hybrids between black and white parents into those of light type and those of dark type, conclude that the former are to the later as 3.1 to 1. Bateson and Punnett (1905, p. 117) conclude that offspring of a pure white parent with colored or heterozygous (mixed) birds are practically always prevailingly white. Hurst (1905, pp. 146-149) gets chiefly white birds from crosses of White Leghorn hens with black or mottled males. The exceptions may be due to the impurity of one of the females.

Barring.—No barring resulted from crossing White Leghorn with Houdan or black Minorca, or Silky with Frizzle. On the other hand, all males, and only males, were barred in the hybrids of Tosa × White Cochin, and in the hybrids of White Leghorn Bantam and Rumpless Game barring occurred, but among males only. Of 26 hybrids between Black Cochin and White Leghorn, 8 were barred black and white, and these belonged equally to the two sexes. Of 11 dark hybrids obtained by Hurst (1905, p. 133) from White Leghorn × Houdan, 6 developed into black females and 5 into cuckoo males. Apparently barring ("cuckoo marking" of the English) is associated with maleness. This result is curious enough, for, as Darwin pointed out, in the ancestors of domestic poultry barring (or rather penciling) is confined to the female sex.

Barring is a heterozygous condition found in hybrids from a white and a black parent. It is provisionally regarded as a form of particulate inheritance as opposed to the alternative inheritance of the Leghorn × Minorca cross. This heterozygous condition when interbred usually breaks up into white, uniformly pigmented, and barred again, as in the case of the Tosa × White Cochin hybrids (p. 49). This form has in certain cases, as in the Cuckoo Dorkings and in the Dominiques—ancestors to the Plymouth Rocks— become truly mosaic, transmitting the mixture of qualities pure. The method of fixing a heterozygous quality is still unknown to science.*

---

* The experience of breeders of mice and guinea-pigs shows that white may be due to the absence of an oxidizing ferment necessary to the bringing out of the color potential in a chromogenic substance (*cf.* von Fürth, 1903). If the chromogen is present the addition (by crossing with a pigmented individual) of the ferment will reveal in the hybrid offspring the colors and pattern latent in the white parent. Working on this hypothesis, we can judge of the latent patterns in the White Leghorn bantams and draw conclusions

Andalusian Coloration.—Among the offspring of a White Leghorn and a Black Minorca two adult blue fowls were reared (fig. 54). The coloration was that of the Andalusian "breed." It consisted of a minute patchwork of black and white pigment. Such a blue coloration is common in barn-yard fowls. It results, according to the testimony of breeders,† from crossing black and white.

The special conditions which determine whether the offspring of a white and a black parent shall be all white or barred or blue have not yet been determined. The solution of this problem offers one of the most interesting fields for future investigation (p. 30).

WHITE vs. BUFF.—Both colors are novel; the former is probably a negative mutation; the latter has been extracted from the original game coloration of fowls. The hybrids are prevailingly white, and white may be regarded as dominant. Nevertheless, this dominance is imperfect, for in half of the offspring buff is more or less evident. It is found diffused over the back, wings, and breast as in "pile" Games. On the whole, white is less strongly dominant over buff than it is over black (Hurst, 1905, p. 134).

BLACK vs. RED.—The red coloration is ancestral; the solid black is novel and positive—a melanic condition. The hybrids between Black Cochin and Red-breasted Game are prevailingly black, but about half of them show red lacing on the hackle feathers or a red peppering in those places where red is displayed by the Game. Black is dominant over red, but imperfectly so.

COLOR OF TOP OF HEAD.

In the white-crested Black Polish the feathers of the top of the head are in striking contrast to those over the rest of the body. That the crest is not necessarily white is proven by the existence of a black-crested race. Hybrids between the Minorca, whose head is wholly black, and the Polish give (p. 15) chiefly black feathers in the males, the females, however, still showing

---

as to what pigmented ancestors they may have had. They were used in five crosses, as follows: (1) Black Cochin × White Leghorn; (2) White Leghorn × Buff Cochin; (3) White Leghorn × Black-breasted Red-backed Game; (4) White Leghorn × Dark Brahma; (5) Dark Brahma × White Leghorn.

Taking all offspring together, about 50 per cent (48.5) are *white* or nearly so. All crosses exhibit barring, together in about one-quarter (26.5) of the cases, and also black and buff or red. It seems probable that all of these pigments and the barred pattern are latent in my White Leghorn bantams. These conclusions are supported by breeding the White Leghorns *inter se*, when, in addition to white offspring, a black and a barred were obtained (p. 40). Similarly among the second hybrids between the Tosa and White Cochin Bantam there appeared a male and a female resembling in plumage coloration the Partridge Cochins (p. 49). This coloration probably lay latent in the gametes of the White Cochin.

† Compare Darwin (1876, I, Chapter VII; 1894, I, p. 270); Wright (1902, pp. 291, 292, 317, 301, 399, 401, etc.); Bateson and Saunders (1902, p. 131); Bateson and Punnett (1905, p. 126). When blues are interbred, the offspring are either white or black or blue. Even in the Andalusian "breed" the blue coloration has never become fixed.

white in their crests. The hybrids crossed back on the Minorca give nearly 100 per cent black heads. Black is dominant, but imperfectly so; the negative characteristic is recessive. The dominant character is less perfectly dominant in the female sex than in the male.

COLOR OF HACKLES—HACKLE LACING.

The color of the hackle feathers and the correlated saddle feathers in birds of broken color usually differs from that of the rest of the plumage. This peculiarity of the hackle coloration is an old character, since it is exhibited by the Jungle fowl, and was probably in the ancestor of the Aseel-Malay group. The feathers are laced with a lighter color than the center.

In crosses between Minorca and Dark Brahma, and White Leghorn and Dark Brahma the solid color (black or white), the new, positive characteristic, dominates over the lacing. Nevertheless, in the Minorca × Dark Brahma hybrids the feathers of the nape are frequently faintly laced with gray. The black is imperfectly dominant.

WING COLOR—RED WING COVERTS.

The male Jungle fowl has red on the upper wing coverts, and doubtless the male of the ancestors of the Aseel-Malay group had also.

The male hybrids between the Dark Brahma and the Black Minorca on the one hand and the White Leghorn on the other usually show red on the wing coverts, although there is no other red in the plumage. Red on the wing coverts is probably dominant, but it is much reduced.

TAIL COLOR.

Although the tail feathers are derived from a distinct feather tract, and in broken-colored birds are usually without the red of the wing, yet tail color does not seem to be a unit character; in inheritance it follows the rest of the body plumage. On the other hand, in breeding buff varieties black persists in the tail feathers longer than in the others. This case resembles the persistence of black at the extremities of the legs of white or red rabbits (Castle, 1905).

SHAFTING.

The female Jungle fowl has a light shaft to the feather. The same is true of the Tosa fowl and some Games. Light shafting is a primitive characteristic of the female.

In the female hybrids between the Tosa fowl and White Cochin the shafting is greatly broadened, and this is the principal modification of the plumage color. In female hybrids of the Tosa fowl and Dark Brahmas the shafting of the feathers of the back and wing coverts is striking, and some shafting appears in two of the males, probably transferred from the female (p. 54). Apparently shafting is dominant.

BODY LACING.

This character is not found in the Jungle fowl, but may have been derived from the penciling of the Aseel-Malay group. In male hybrids between the Tosa fowl and the Dark Brahma it occurs, derived from the latter (p. 54). It appears to be dominant.

PENCILING.

This is an ancient feminine characteristic, best marked in the Aseel-Indian group (p. 53). It is found particularly well developed in the Dark Brahma female. In the female hybrids between that race and the Tosa fowl penciling is well developed; it is dominant.

GENERAL TOPICS IN INHERITANCE.

UNIT CHARACTERS.

Taxonomic descriptions of plants and animals give a list of their specific characteristics (Merkmale, caractères). These comprise for the most part only the external characteristics, but a similar list might be made for internal characteristics. In addition to specific characteristics, those of a higher order (such as generic, etc.) and those of a lower order (such as varietal) may be enumerated. Such characteristics are, in first approximation, unit characters. They are of prime importance, because the whole problem of evolution is that of the origin and significance of the various unit characters of the body.

The theory of the unit character is associated with that of its bearer in inheritance. Darwin (1876) and later de Vries (1889) designated as such bearers particles of the nuclear material named "pangenes." "Changed numerical relation of pangenes is the basis of fluctuating variability; displacement (Umlagerung) of pangenes in the nucleus conditions retrogressive and degressive mutations; while the formation of new kinds of pangenes is necessary to the explanation of progressive mutations" (*i. e.*, those exhibiting altogether new characteristics).

The two main hypotheses of the origin of unit character are that of de Vries and that of Weismann. De Vries sets forth his hypothesis at the very beginning of his great work, "Die Mutationstheorie." His words may be thus translated:

As mutation theory I designate the doctrine that the characteristics of organisms are built up of units that are sharply separable one from another. These units can be united into groups, and in related species the same units and groups recur. Transitions, such as the external forms of plants and animals exhibit in such numbers, exist between the units as little as between the molecules of chemistry. . . . . In the realm of the doctrine of descent this principle leads to the conviction that species have proceeded from one another not continuously but by steps [nicht fliessend, aber stufenweise]. Each new unit added to the older ones constitutes a step and separates the new form, as an independent species, sharply and fully from the species whence it arose. . . . . Die neue Art ist somit mit einem Male da; sie entsteht aus der früheren ohne sichtbare Vorbereitung, ohne Übergänge.

Weismann, on the other hand, is only less clear in expressing his hypothesis. He accepts, of course, the idea of unit characters, each of which is represented in the germ cells by a "determinant." "We called," he says (1904, I, p. 369), "determinants those parts of the germ-substance which determine an 'hereditary character' of the body; that is, whose presence in the germ determines that a particular part of the body, whether it consists of a group of cells, a single cell, or a part of a cell, shall develop in a specific manner, and whose variations cause the variations of these particular parts alone." The "hereditary parts" may be small or "large regions, whole cell masses of the body, which in all probability vary only *en bloc*, as, for instance, the milliards of blood cells in man, the hundreds of thousands or millions of cells in the liver and other glandular organs, the thousands of fibers in a muscle, or of the sinews or fascia, the cells of a cartilage or a bone, and so on. In all these cases a single determinant, or at least a few in the germ plasm, may be enough." For Weismann (1904, II, p. 151) the ultimate source of all hereditary variations is the variation of the representatives of the unit characters in the germ plasm. "If I mistake not," he says, "we may at least say so much, that all variations are, in ultimate instance, quantitative and that they depend on the increase or decrease of the vital particles, or their constituents, the molecules. . . . . What appears to us a qualitative variation is, in reality, nothing more than a greater or less different mingling of the constituents which make up the higher unit; an unequal increase or decrease of these constituents, the lower units." The cell changes its *constitution* when the proportion of its component parts "is disturbed, when, for instance, the red pigment granules which were formerly present, but scarcely visible, increase so that the cell looks red. If there had previously been no red granules present, they might have arisen through the breaking up of certain other particles—of protoplasm, for instance, in the course of metabolism—so that, among other substances, red granules of uric acid or some other red stuff were produced. In this case, also, the qualitative change would depend on an increase or decrease of certain simpler molecules and atoms constituting the protoplasm-molecule."

In criticism of the foregoing it may be said that a variation in the number of atoms in a protoplasmic molecule is certainly also a qualitative change—a mutation. The only real difference between Weismann and de Vries depends on the extent of the mutative modification, whether progressive or complete from the beginning; but this is a real difference, for the latter view is required by the theory of immutable unit characters. The former view is not in harmony with such a theory. Conversely, if it appears that there are immutable unit characters, then the theory of evolution by saltation is necessary; if unit characters are modifiable, then species may have arisen gradually.

The result of the breeding experiments described herein bears upon this discussion. No other group, I imagine, exhibits so many characteristics as poultry ; of the comb alone there are half a dozen forms. The forms of feathers and their color patterns are numerous. These forms are sharply marked off from one another for the most part ; moreover, when two characteristics are crossed the result is rarely a blend. This was a great surprise to me, as I had anticipated that blends would be the rule ; and, overwhelmed by the facts, I embraced at once the theory of immutable characteristics.

That there are unit characters in poultry can not be doubted. When single and V comb are crossed and progeny obtained all with a Y comb, how convincingly do the second hybrids reproduce the single comb in some individuals and the V comb in others! Though the cerebral hernia and its associated great crest may disappear in the first generation of hybrids, how beautifully do they reappear in one-fourth of the offspring of such hybrids! How instructive is it to see perfectly plain feathered offspring arising from a frizzled pair, or in a Black Minorca × Dark Brahma white-laced hackles appearing in an otherwise dead-black plumage! Truly we may hope, as in chemistry, to make various kinds of molecules by the proper admixture of our atoms—the characteristics. Even in man such non-blending characteristics are evident. One of the most famous is the Hapsburg lip or chin, which from the fifteenth century has persisted to the present day despite infusion of new blood during fifteen generations.* Another striking case is that of hypophalangia in man, described by Farabee (1905). In the four or five generations studied, there has, he states, "never been a single instance of partial inheritance, but in all cases all extremities have been affected in precisely the same way."

While admitting, thus, the reality of unit characters, the further study of the evidence of hybridization in poultry has led me away from the conception that they are rigid and immutable as atoms are, which may be combined and recombined in various way and always come out of the process in their pristine purity. This is by no means the case. Very frequently, if not always, the character that has been once crossed has been affected by its opposite with which it was mated and whose place it has taken in the hybrid. It may be extracted therefrom to use in a new combination, but it will be found to be altered. This we have seen to be true for almost every characteristic sufficiently studied—for the comb form, the nostril form, cerebral hernia, crest, muff, tail length, vulture hock, foot-feathering, foot color, earlobe, and both general and special plumage color. Everywhere unit characters are changed by hybridizing.

How does this fact bear on the rival theories of evolution? It has an important bearing on them. It is not in accord with the statements of de Vries

---

* *Cf.* F. A. Woods, 1902–03.

quoted above: "The characteristics of organisms are built up of units that are sharply separable one from another," and "Transitions exist between the unit as little as between the molecules." Single comb is one unit and pea comb is a different unit, but they are not sharply separable. Crest and no crest are units, but they run into each other in hybridizing. Unit characters may show transitions, and, if so, they *may* have originated gradually, so far as I see. It does not follow that they must have originated gradually.

ALTERNATIVE, PARTICULATE (MOSAIC), AND BLENDING INHERITANCE.

Doubtless Darwin's statement that crossed characters usually blend is still the prevalent view. Much if not most biometric work in heredity has been made on this basal assumption. I may say that I began my experiments prejudiced in favor of this view.

The results that have been recorded in the foregoing pages indicate that probably in general typical blending of characters is rare. Excepting characters like general form of the body, which are doubtless not units, but complex, I have, indeed, seen no single case of a typical blend. A fusion of characters is a rather rare phenomenon. Human skin color is the one striking case. One can but wish we had more careful data on inheritance of human skin color in successive generations. Other human characteristics show alternative inheritance. This is strikingly true in Farabee's family of hypodactyls cited above. It is said to be true of eye color and probably of the states of general pigmentation known as blonde and brunette.

The following characters of poultry show alternative inheritance:

Comb form.	Uropygium.	Earlobe color.
Nostril form.	Tail length.	General plumage color
Cerebral hernia.	Vulture hock.	(sometimes).
Crest.	Booting.	Color of hackles.
Muff.	Extra toe.	Wing bar.
Beard.	Color of mandible	Shafting.
Frizzling.	and foot.	Body lacing.
Silkiness.	Iris color.	Penciling.

The following characteristics show particulate inheritance:

Iris color (sometimes?).
White and black, producing *barring* (Series VI, VII, IX, XII).
White and black, producing blue (fig. 54) (a fine mosaic of white and black).

It is too early yet to interpret the cases of particulate inheritance. It is a striking fact that, excepting the Tosa × White Cochin cross, all my barred birds reared to maturity had the White Leghorn Bantams as mother or father. Now, as repeatedly observed, these bantams were heterogametous. It is possible that they contain barred blood in the "fixed" condition. Aside from the fact that they throw a certain proportion of barred birds, this conclusion gains support from the fact that the wing coverts of the male are obscurely barred with dusty bands, although, on the other hand, this barring may be merely the badge of heterozygotism. It is possible, therefore, that the barring in the plumage of the White Leghorn Bantams is transmitted as an alternative characteristic. The case of the barred descendants of the Tosa × White Cochin is more difficult. I am not yet prepared to go so far as Correns (1905[4], p. 13, note) when he says: "Wo Mosaikbildung als *Regel* bei einem Bastard auftritt, war sie schon in einem der Eltern oder in beiden, aktiv oder latent, vorhanden."

Naturally, attention was directed chiefly toward evident qualitatively marked characteristics. Such do not blend. The fact that for the most part a characteristic does not blend when crossed with its allelomorph is of the highest importance for the theory of evolution. If blending were universal a new characteristic must inevitably become quickly swamped by intercrossing with the parental characteristic. Since the new quality does not blend, it need not be swamped, even when there is no special isolation.

INHERITANCE OF SPECIFIC VS. VARIETAL CHARACTERISTICS.

A distinction between specific and varietal characteristics is made by Nägeli (1884, p. 247) and by de Vries (1902, p. 141 ; 1905, p. 141). Following de Vries, a specific characteristic is a wholly novel one acquired by the race—one which stamps its possessor as an elementary species. A varietal characteristic is sometimes positive (*i. e.*, additional), in which case it is found also in closely allied species, and may be regarded as the becoming patent of a characteristic all the time latent in the variety. It is, on the other hand, sometimes negative, this condition being marked by the disappearance (becoming latent) of a characteristic patent in the ancestral species. Specific and varietal characteristics are thought by de Vries to be inherited very differently. When two elementary species are crossed the characteristics of both parents appear, fully developed, side by side ; Mendel's law is not followed. When a species is crossed with a variety a Mendelian result is obtained and the patent characteristic is dominant over the latent.

Let us now see in how far the results gained in breeding poultry accord with de Vries's law. It is not easy to make the classification in an unprejudiced way ; an attempt, however, will be made.

First, the comb is a specific characteristic of the genus *Gallus*. It is absent in other Gallinæ. Also pea comb and rose comb are each wholly new, positive variations from the primitive single comb. Muff and beard seem to be novel ; so also the long tail of the Tosa fowl, the extra toe, and the melanic feet and beak.

Clear cases of negative variations are: Loss of the nasal process of intermaxillary and consequent *high nostril;* failure of cerebral plate to close and consequent cerebral hernia ; loss of uropygium ; loss of red and black pigment in feathers (albinism, partial or complete) ; loss of dark pigment in crest feathers ; loss of wing bar ; loss of primitive shafting.

Now, by hypothesis we should expect a difference in inheritance in these characteristics as indicated below :

Expected non-Mendelian.	Expected Mendelian.
1. Pea and rose comb *vs.* single comb.	1. High nostril *vs.* low.
2. Muff and beard *vs.* plain head.	2. Cerebral hernia *vs.* normal.
3. Long tail *vs.* normal.	3. Taillessness *vs.* normal tail.
4. Extra toe *vs.* normal.	4. Albinism in plumage *vs.* pigment.
5. Melanic feet and beak *vs.* willow or yellow.	5. Absence of wing bar *vs.* presence.
	6. Absence of shafting *vs.* presence.

Of the five cases where, on de Vries's theory, we should expect non-Mendelian results, No. 4 is apparently not Mendelian, No. 5 gives often a mixture of characters, Nos. 1 and 2 apparently give true Mendelian dominance and recessiveness, No. 3 is still doubtful. Of the six cases in which a Mendelian inheritance is looked for, we certainly find it in three cases and less certainly in the others. On the whole, there is a slight but not a striking difference in transmission between the two sets of characteristics, and I can only conclude that for poultry, so far as I can see at present, de Vries's formula does not hold universally.

INHERITANCE OF POSITIVE VS. NEGATIVE VARIETAL CHARACTERISTICS.

According to de Vries, when an individual having a certain characteristic patent is crossed with one in which it is latent the patent characteristic is dominant, the latent recessive. Do results with poultry confirm this law?

In the following table the patent characteristic is given in the left-hand column and the dominant characteristic in *italics* :

Patent.	Latent.
1. Nasal process of premaxillary, *narrow nostril.*	High nostril.
2. Closure of cerebrum completed ; *plain head.*	Failure of cerebrum to close; cerebral hernia.
3. Crest ; *black crest feathers*............	Smooth head ; white crest feathers.
4. *Complete development of the feather*..	Interrupted development of the feather; silky feather.
5. *Tail.*	Taillessness.
6. Pigmented plumage	*Albinism* in plumage.
7. *Red wing bar.*.	Uniformly colored wing.
8. *Shafting.*	Plain feather.

Of the foregoing eight characters, seven clearly follow the law that patent characteristics dominate over latent. No. 6 is a clear exception, for since all the wild Gallinæ are deeply pigmented birds it can hardly be doubted that white is a negative variation in which color is latent. However, the exception (No. 6) is not universal, for white plumage does not always dominate over pigmented plumage. It appears, then, that the patent character is, in general, but with some exceptions, dominant over the corresponding latent character.

INHERITANCE OF OLD VS. NEW CHARACTERISTICS.

Standfuss (1896, p. 111), as a result of his hybridization of moths, concluded that hybrids resemble the older species. De Vries (1902, pp. 33-42, and 1905, pp. 280, 281) cites several instances of the prepotency of the phylogenetically older characteristic. Bateson and Saunders (1902, p. 137), however, point out that younger characteristics sometimes dominate, and cite pea and rose comb, extra toe, and the polled condition of cattle as examples. Correns (1905, p. 463 et seq.) describes a case of petaloid calyx—a new characteristic—which is dominant over the normal form. Correns (1905^a, p. 13), indeed, concludes that in general the phylogenetically more advanced characteristic—the later originated, *younger* characteristic—dominates.

Let us see what evidence poultry hybrids have to offer bearing on this point.

Old characteristics.	New characteristics.	Old characteristics.	New characteristics.
1. Single comb.....	*Rose comb*.	11. White skin.....	*Black skin*.
2. *Low nostril*.....	High nostril.	12. Red iris........	*Black iris*.
3. *No hernia*......	Hernia.	13. *Red earlobe*....	White earlobe.
4. Plain head......	*Crest*.	14. Pigmented....	*White* (sometimes dominates).
5. No muffling.....	*Muffling*.		
6. Plain feathers...	*Frizzled feathers*.	15. Red pigmented.	*Black; no red*.
7. *Plain feathers*...	Silky feathers.	16. *Black head*.....	White head.
8. *Tailed* (?).......	Non-tailed.	17. Hackle lacing..	*Solid black*.
9. Tail feathers limited in growth.	*Tail unlimited*.	18. *Red wing bar*..	No wing bar.
		19. *Shafting*.......	No shafting.
10. Four toes........	Five toes.	20. *Penciling*......	No penciling.

This table shows that of nineteen characteristics (No. 10 being left out of consideration), nine old ones are dominant and ten new ones. Clearly, dominance of characteristics in poultry is not determined by the age of the characteristic.

DOMINANCE AND RECESSIVENESS.

Mendelian dominance and recessiveness with segregation of characteristics in the gametes are not universal concomitants of hybridization. Mendel knew it (Correns, 1905^a); de Vries founds his system on the fact; Correns lays stress on it; Bateson and Saunders (1902, p. 152) recognize it, but consider the exceptions insufficiently known. The characteristics that I have crossed show always segregation excepting extra toe and perhaps also melanic foot and beak color. These are among the positive variations of de Vries, which, in accordance with his system, we should not expect to "mendelize." As stated, other positive variations, however (pea comb and muff), seem to mendelize.

Of the varietal characteristics, the positive or patent characteristics almost always are dominant, white plumage forming an occasional exception. On the other hand, phylogenetically old characters are not more apt to be dominant than "new" ones. Some evident sports, such as crest, frizzling of feathers, unlimited growth of tail, and black skin (of Silky), are dominant. Other sports—hernia, shortened premaxillary, silkiness, and rumplessness—are recessive; the novelty or antiquity of the characteristic has nothing to do with its dominance. Dominance of a character in hybridization is determined by the same causes as determine the appearance in the race of a positive variation. A progressive variation, one which means a further stage in ontogeny, will be dominant; a variation that is due to abbreviation of the ontogenetic process, which depends on something having dropped out, will be recessive.

This conclusion, however satisfactory, must be regarded as tentative. It is doubtful if it is of general validity; for while long tail and crest feathers are dominant in poultry, long hair (equally due to prolonged life of the follicle) is recessive in mammals (Castle, 1903; 1905, pp. 64–67, 73–74; Hurst, 1904). White is usually recessive to pigment in flowers and mammals, but it is usually dominant over pigment in poultry. It is still too early to regard the conclusions expressed in the last paragraph as anything but an hypothesis.

While dominance and recessiveness are typically found in Mendelian inheritance, yet *they may be absent* even in cases when segregation of characteristics occurs in the second hybrid generation. Thus, the barred offspring of the black-and-red Tosa fowl and the white Cochin throw in the F_2 generation 25 per cent black and red and 25 per cent white, but the remainder, like all of F_1, is barred with white, and no one can say which plumage color is dominant. The same is true of some black-and-white barred hybrids. It is also true of hybrids between single and V comb. The phenomena of dominance and recessiveness do not always accompany segregation.

Another modification of the law of dominance and recessiveness must be recognized, namely, that they are by no means always complete. Even in the first hybrid generation the dominant characteristic is more or less intermediate. The antagonistic characteristics a and a' of the two parent types are not only united in the zygote, but they pass in the development of the organism into all the tissues of the body, and particularly into the cells out of which the organ A is developed. The dominant characteristic, a, and the recessive characteristic, a', each works to determine the quality of the organ A. If a dominates, it is because it is more active than a'. It does not dominate by excluding a'. Sometimes, as in the case of barred feathers, it appears that a and a' in ontogeny alternate in their activities. The cells of a certain zone of the feather manufacture only black pigment; in the next zone black is wanting; then comes a zone of black, and so on, in many repetitions. Dominance as contrasted with recessiveness is a matter of degree and not of kind.

Various authors refer to the imperfection of the dominant or recessive characteristic in the hybrid. Bateson and Saunders (1902, p. 23) say:

Although the offspring resulting from a cross between any two of the forms (of *Datura*) employed is usually indistinguishable from the type which is dominant as regards the particular character crossed, yet in other cases the intensity of a dominant character may be more or less diminished either in particular individuals or in particular parts of one individual.

Hurst (1905, pp. 145-154) records many cases of imperfect dominance in poultry and estimates the incomplete dominants to be twice as numerous as the complete dominants.

DEPENDENCE OF DOMINANCE ON THE RACES CROSSED.

Is one of a pair of allelomorphs that shows itself dominant when varieties *A* and *B* are crossed likewise dominant when any other varieties, *M* and *N*, are crossed, or is the relative potency of the allelomorphs dependent upon the varieties in which they happen to reside?

Data for an answer to this question are to be found in the experiments where the same pair of allelomorphs were crossed, using different varieties. We may except from this list the Minorca × Polish and the Leghorn × Houdan crosses, as the races involved are very closely related. The following allelomorphs remain for consideration:

(1) Crest *vs.* crestlessness.
(2) Silkiness *vs.* non-silkiness.
(3) Rumplessness *vs.* tail.
(4) Vulture hock *vs.* plain hock.
(5) Boot *vs.* clean foot.
(6) Extra *vs.* normal toes.
(7) Black *vs.* white skin.
(8) Black *vs.* yellow beak.
(9) White *vs.* dark plumage.

Crest.—This is dominant when Polish or Houdan is crossed with the Mediterranean breeds and when the Silky is crossed with the Frizzle or with the Jungle fowl. Crest is uniformly dominant over crestlessness, no matter which of these races are used.

Silkiness is recessive to non-silkiness when crossed with Frizzle or the Jungle fowl. Non-silkiness is probably always dominant.

Rumplessness in a Game fowl was recessive to the tailed condition of Leghorn, Cochin, Frizzle, and Nankin. The tailed condition seems always to dominate.

Vulture hock is recessive when an Asiatic race is crossed with any Mediterranean breed or a Game, and probably, in general, plain hock dominates.

Booting is dominant when the booted form is the mother, no matter what the race. Booting is much reduced and sometimes altogether absent in the first generation of hybrids when it is derived from the father. Inheritance of booting is independent of race but not of sex (p. 38).

Extra toe seems not to Mendelize. The excess of extra toes in the first hybrid generation holds for all the races crossed and is probably independent of race.

Black skin of the Silky dominates over the colorless skin of the Frizzle and of the Jungle. It probably dominates throughout.

Yellow vs. black beak and foot color.—Yellow of the White Leghorn dominates over black of the Minorca, but yellow of the Dark Brahma is dominated by the Minorca. Here yellow behaves differently, according as it is in the Leghorn or Dark Brahma race. It is quite possible that the yellow is not identical in the two groups, but that, while it is ancestral in the Dark Brahma, it is secondary and a progressive character in the Leghorn. The lack of uniformity in dominance of yellow may be due to essential dissimilarity of the character in different races.

White vs. dark plumage.—Aside from cases of barring and Andalusian coloration, white usually dominates over dark plumage. This is true in all cases where White Leghorn is employed as white race, whether the other race is Game, Dark Brahma, Houdan, or Minorca. When the Silky is used as the white race white is sometimes recessive (fig. 53), but it must be acknowledged that the dark parents were not the same as were used with the Leghorn, but were a Game, Frizzle, and Jungle fowl; consequently the results in the two series are not strictly comparable. However, Darwin found the white of the Silky recessive to the black of the Minorca. It is hardly conceivable that the white of the Silky is different from that of the Leghorn; so it must be concluded that white inherited as a solid color is sometimes dominant and sometimes recessive, depending on the race in which it inheres.

Summarizing the foregoing evidence, it appears, first, that (except in certain obviously complex color characters) when one of a pair of allelomorphs is dominant it is so regardless of the races crossed. This shows that dominance and recessiveness depend upon a relation of the characteristics *per se* and not upon any relation of the races into which they have been introduced. This is in accord with the conclusion reached above, that dominance is determined by the *positive nature of the characteristic* (p. 84).

PREPOTENCY AND DOMINANCE.

Prepotency was a much used and probably abused term in the period preceding the revival of Mendelism. In the new era all of the old terms have been subjected to reëxamination as to their significance. Bateson and Saunders (1902, p. 121) use the term "as signifying determination of dominance," *i. e.*, whether the normally dominant or the normally recessive character shall be in any case actually dominant. Castle (1905, pp. 58–64) shows that although rough coat is dominant over smooth coat, a few smooth-coated mothers will, when crossed with rough males, produce *partial-rough* young. The normally recessive character here partially dominates. In my own experiments the most remarkable case of dominance is exhibited by a gamete from the maternal side that produced the Houdan × White Leghorn hybrid

No. 386 ♀; for this hybrid has a high nostril and a pair of papillæ like the Houdan mother, both of which characters are recessive. Out of 41 individuals No. 386 is the only one that exhibits them. It appears, then, that "prepotency" in its modern sense can not be neglected.

HYBRID FORMS.

It sometimes happens when two dissimilar characteristics are crossed that neither appears in the offspring, but they are replaced by a new character. This fact has been long known. Mendel obtained such hybrid forms (*cf.*, Correns, 1905, p. 232). Several cases are cited by Focke (1881, pp. 473, 474). He refers particularly to the blue hybrid of the white *Datura ferox* crossed with the likewise white *D. lacvis* and *D. strammonium Bertolonii*.

As a result of more recent work it appears probable that hybrid forms are of two kinds. First, such as are atavistic or due to the becoming patent of a latent characteristic;* and, second, such as are due to a particulate inheritance of the two characteristics crossed. In the latter case all that is novel in the hybrid is the replacement of either *single* character by a combination of characteristics.

Atavistic hybrid forms have been carefully investigated of late, especially by Correns (1902) and Cuénot (1903), who have applied a method of interpretation to particular cases. When albino mice are crossed *inter se* they produce only albinos. But if such an albino is crossed with a pigmented (*e. g.*, a black) mouse its latent pigment appears and the offspring may be all gray, or perhaps yellow and gray or yellow and black. The same holds exactly true for albino rabbits, as Hurst (1905, pp. 306–310) has shown. Cuénot's interpretation depends on the principle that pigments result from the action of an oxidizing diastase (tryosinase) upon a chromogenic substance. Both of these elements are present in a pigmented mouse, but he assumes the chromogenic substance alone is present in the albino. The sperm from the pigmented male brings to the egg of an albino the diastase necessary to the production of pigment in the offspring. Correns (1905°) finds that the hybrid of *Mirabilis jalapa alba* (white flowers) and *M. jalapa gilva* (yellow flowers) has rose-colored flowers that are, moreover, *striped* with red. His experiments lead to the conclusion that the *alba* variety forms no pigment, but does produce a pigment-changing (reddening) enzyme. The *gilva* variety forms pigment, but not the reddening enzyme. When *alba* sperm unites with the *gilva* egg the pigment of the latter, under the influence of the reddening enzyme, becomes rose. Similarly with striping. There is evidence that this is only partly latent in *alba* and completely latent in *gilva*. Now if we assume a factor that permits the development of the striping determinant to be active in *gilva* but to be latent in *alba*, the imperfect

---

* Tschermak (1904, p. 95) would add as another kind that in which an originally patent character becomes latent.

development in *alba* of the striping determinant is accounted for. When the sperm of *gilva*, bringing the active principle for striping, fertilizes the egg of *alba* with its striping determinant, the striping makes its full appearance. These two or three examples from both plant and animals indicate a method of explaining hybrid forms that is probably of wide applicability.

Are the hybrid forms of poultry to be explained on the atavistic or the particulate inheritance theory? Take first the case of barring. Three tests can be applied: First, inherent probability from the ancestry of the fowl crossed; second, general distribution of barring among the offspring; third, proportion of different forms of plumage pattern in generations beyond the first. The cross between Tosa fowl and White Cochin gave barred birds. If the barring were latent it must have lain in the Cochins—the form without visible pattern. It is fairly certain that neither of the ancestors of domestic fowl was barred; hence if the barring determinant existed in the Cochin bantam it must have been introduced by a recent cross. Bantamizing of Cochins is effected by crossing with some bantam race, but until recently no barred bantams have been created. It is therefore highly improbable that a barred bird was used to bantamize the Cochins. While it is possible, it is improbable that the White Cochin contained a barred determinant. Second, barred *races* have the two sexes equally barred, but our hybrids are barred in the male only; consequently barring here acts like a neomorph. Third, on the theory of atavism we should expect to get in the second hybrid generation:

Coloration of second generation.	Atavism theory.	Particulate theory.	Actual.
	Per cent.	*Per cent.*	*Per cent.*
White	25.00	25	28
Pigmented and barred	56.25	50	48
Pigmented and not barred	18.75	25	24

The actual proportions of the three types accord much better with the particulate inheritance theory than with that of atavism, but the total number of offspring is insufficient to give certainty. It may be concluded that while the evidence does not exclude the atavism theory of the cropping out of barring, it favors the theory of particulate inheritance.

The case of the hybrid between single and V comb rests on more extensive data. These are set forth on pages 10–12, and are less favorable to the atavistic theory than to the particulate theory.

The other heterozygous forms have been less carefully studied. They are the blue, Andalusian (fig. 54, pl. XVII), plumage color resulting from a white and a black crossed, and the case of the down of the hybrid Minorca × Dark Brahma chicks. This is black like the Minorca, but lacks the white of the chicks both of that race and of the Dark Brahma. The Andalusian breed has been discussed by Bateson and Punnett (1905, p. 126), and they find, what is the universal testimony of breeders, that (as stated also at page 76) the blues bred *inter se* produce some white and some blacks, but still more blues. Until more complete statistics have been gained on the proportions of colors in the offspring, the interpretation of blue must remain uncertain.

Hybrid forms are, then, frequently cases of particulate inheritance in which the hybrid gametes are not mosaic; consequently whenever "pure" offspring are produced, as in F_2, these reassume the character of the pure race. In some cases, as in the cuckoo Dorking and the Dominique (from which our barred Plymouth Rock has been derived), the heterozygous form of barred plumage has become fixed, so that only barred offspring are produced. A mosaic gamete has been created. The blue coloration has never yet been fixed as a permanent hybrid form. The method of fixing a hybrid form is urgently in need of investigation.

REVERSION.

This term has been used rather loosely in the past for the appearance in hybrids of characteristics not visible in the immediate parents of the hybrids and often belonging to remote ancestors. Darwin (1876) made much use of this term in describing his results. He believed that the occurrence of "reversion" gave a useful key to ancestry. It is worth while to consider his observations and experiments. He mentions the fact that "purely bred Game, Malay, Cochin, Dorking, Bantam, and . . . . . Silk fowls may frequently or occasionally be met with, which are almost identical in plumage with the wild *G. bankiva*." But does this indicate anything else than that this type of coloration has persisted in certain primitive races, like the Game, and has been transplanted from them to the new races? Darwin crossed a black Spanish cock with various white and white-and-black hens of pure breed. The offspring of this cock crossed with a silver-spangled Polish hen and with a white Cochin hen showed no sign of reversion to the red color of *G. bankiva*. The male offspring of a spangled or silver Hamburgh hen showed white in the hackles and a reddish yellow on the saddle. Darwin regarded this as a "first symptom of reversion;" but in the first of these peculiarities the hybrid resembles *G. bankiva* less than the Dark Brahma. The offspring of a white Game hen with the Spanish cock was at first snow white, but eventually produced the "pile" coloration. Darwin regards this as a partial reversion to *G. bankiva;* but it is equally possible that the reversion is only to a pile coloration that is latent in the white from an earlier cross and is brought out when the white is crossed with a dark color. But Darwin's most remarkable hybrid was the offspring of a white Silky hen. Of two cockerels one was black (with light laced hackles); the other resembled closely a Jungle cock. Darwin admits that the case is extraordinary, but it was duplicated by Mr. Tegetmeier. This experiment certainly should be repeated, and I have arranged to repeat it next season.

One of the best cases of reversion is the gray coat of a hybrid between a white and a black mouse. We now know, however, that even a "pure race" of white mice may carry gray as a latent characteristic that first becomes patent on crossing. In view of such facts cases of "reversion" to a *remote* ancestor must be critically examined. If the "reversion" be not a neomorph, it must have been handed down without break in the germ plasm from an ancestor possessing the characteristic.

PURITY OF GAMETES.

The dogma of purity of the gametes, the second corner-stone of Mendelism, asserts that while the unripe germ cells of a hybrid having antagonistic or alternative characteristics A and A' contain representatives of both A and A', yet the *ripe* germ cells of such a hybrid contain representatives of either A or A', and not of both. Thus the ripe germ cells (gametes) are pure in respect to a given characteristic. They gain this purity, it is supposed, during the maturation period, the period when the reduction division of the chromosomes occurs, and when in each cell division one-half of each chromosome moves bodily to one daughter cell and one-half to the other. The theory assumes, of course, that characteristics A and A', being derived from different parents, inhere in different chromosomes. Let us assume that our hybrid has eight chromosomes, four derived from each parent, thus:

in which the black dots represent chromosomes of maternal origin; the circles chromosomes of paternal origin. If all maternal chromosomes contain the determinant a then purity of the gametes demands that all such go to one gamete and all of the chromosomes of paternal origin go to the other, and that such is their behavior has in fact been assumed by Cannon (1902). But that would result in the extracted pure individuals of the second hybrid generation being like their grandmother or their grandfather in all characteristics, which is not the case. If we assume that some only of the maternal chromosomes, such as are represented by the small dots, contain the determinant a, then these may be associated with any of the paternal chromosomes *excepting* those that contain the determinant a'. Such a selection of chromosomes so as to exclude from the ripe gamete chromosomes containing both the alternative characteristics is quite possible, owing to the fact of synapsis, in which the *homologous* chromosomes from the two parents unite in pairs, as shown in the figure, in such a way that both can not pass to the same gamete.

The foregoing hypothesis of Sutton (1902, 1903) and Boveri (1902) would account for perfect purity of gametes. But it is clear that gametes are not wholly pure, since the characteristics in second generation hybrids are rarely exactly like those of their grandparents; consequently various additional hypotheses have been offered accounting for this feature. Häcker (1904) points out that chromosomes do not pass from cell to cell unchanged except for their growth and division. During the resting stage "of the nucleus it disappears. The new nucleus which arises in the position of the old is at first small; it arises inside of the old chromosome as a spore arises in the mother cell; its material has been derived from a part only of that of the mother chromosome; the remainder goes to form part of the cytoplasm. Though chromosomes from different parents tend to separate to distinct gametes, still all gametes are infected by each kind of characteristic." McClung (1905, p. 329) assumes, more vaguely, a mutual influence of synaptically paired chromosomes in the prophase of the first spermatocyte.

A different suggestion is offered by Ziegler (1905). He assumes that each chromosome of maternal or of paternal origin carries determinants of *all* characteristics. After maturation all gametes contain the same number of chromosomes, but the proportion in them of chromosomes of paternal and of maternal origin varies. Gametes rarely contain exclusively maternal or paternal chromosomes, but whenever the proportion from one parent is high the gamete acts as though it contains exclusively the gametes of that one ancestor. If two gametes that are prevailingly paternal unite in a zygote the resulting hybrids (of the second generation) show all the grand-paternal characteristics. The difficulty with this hypothesis is that, like Cannon's, it does not account (any better than the first hypothesis) for the diverse combinations of characteristics shown in the second hybrid generation.

Still another suggestion has been made by Morgan (1905). It is that the gametes are not pure, but contain determinants of both allelomorphs a and a', and that one of these dominates in half of the gametes and the other in the remaining half. The advantage of this hypothesis is that it accounts for latent dominant characters in recessive individuals. This hypothesis assumes that the gametes of hybrids are always impure, and that this impurity can not be got rid of. This seems to me to be contrary to experience. Moreover, except for the explanation that it offers of latency—which has been accounted for on other grounds by Cuénot—it offers no practical advantage over the theory of pure gametes.

From the foregoing diversity of hypotheses it is evident that we lack a fully satisfactory cytological explanation of the facts other than that of purity—the fact of imperfect dominance and the fact of particulate inheritance combined with purity in the second hybrid generation. Perhaps it will suffice to suppose a restricted purity of gametes such that the determinant of a characteristic may become infected to a slight degree by the presence of its allelomorph.

COMPARISON OF RECIPROCAL CROSSES.

There is a notion among breeders of poultry that the father and the mother contribute different qualities to the offspring; and if the cytoplasm carries any hereditary tendencies this result is to be expected, for the female transmits more cytoplasm than the male. Certainly the hybrid between a large hen and a bantam cock starts life on a very different plane of size from the hybrid between a bantam hen and a large cock. A writer in Wright's Poultry Book (1902) says in respect to breeding Houdans that the male bird is more responsible for the outside qualities—color, size of crest, beard, tail carriage, color of legs, and so on. The hen determines laying qualities and general size.

I have made only one extensive experiment on this matter. I crossed a single-comb White Leghorn bantam and a Dark Brahma both ways. The offspring of the Dark Brahma hen (weight, 1,300 grams) are a little heavier than those of the White Leghorn bantam hen (weight, 700 grams). Two males descended from the one and the other mother, respectively, weighed at 3½ months 720 and 550 grams. The average of three pullets from the Dark Brahma at 3 months 22 days is 655 grams; of three pullets from the White Leghorn at 3 months 23 days is 626 grams. The proportional difference in the weight of the young of about 3 to 4 months is less than that of their parents, but is in the same sense.

The booting of the offspring of the White Leghorn hen is much reduced as compared with the booting of the offspring of the Dark Brahma hen, the father in the first cross not differing from the mother in the second cross in its heavy booting. In plumage color the 19 offspring of the White Leghorn female were all white except four. Of the 19 offspring of the Dark Brahma female, only six were white, the others resembling the Dark Brahma. Thus we see that in these three characters of weight, booting, and plumage color the offspring tended to "take after" the mother.

INHERITANCE OF SEXUALLY DIMORPHIC CHARACTERISTICS AND SEXUAL DIMORPHISM IN THE HYBRIDS.

Most species of vertebrates exhibit certain characteristics peculiar to one or the other sex, and it is well known that, for example, a female peculiarity can be transmitted through a son to a granddaughter. Thus the good milking quality of a cow is transmitted through her son to his daughters. Whenever femaleness crops out in the history of the germ plasm the good milking quality, or whatever other quality it may be, also appears. The inheritance of dimorphic characters is most strikingly seen in hybridization. Thus I crossed a male Tosa fowl (which has self-colored feathers) with a white Cochin.* The male hybrids are barred with white, but the female hybrids closely resemble in color the female Tosa fowl in having white

---

*See Series IX.

shafting on the contour feathers, although the white shafting is much broadened. When the barred male and broad-shafted females of this first hybrid generation were crossed the pure plumage of the Tosa fowl tends to reappear. The males have contour feathers without white and with much red; the females have the shafted feather without any red. With maleness or femaleness go the proper secondary attributes.

What is true of the Tosa fowl is true generally, and there is much opportunity to test this matter in poultry, for sexual dimorphism is widespread. In all "dark" or "partridge," silvered, and golden races as found in Brahmas, Cochins, Wyandottes, Dorkings, Hamburghs, Games, and Oriental fowl, the plumage of the two sexes is conspicuously different; and to produce sexual dimorphism in a race that is without it the use of one male bird of a dimorphic race may suffice. Again, in the male, comb and wattles are generally larger than in the female. The rose comb of the male becomes often a modified pea comb in the female. The simple comb of the Minorca, Spanish, and Dorking fowl is erect in the male, drooping to one side (equally to the right and left side) in the female. The form of the hackle and saddle feathers constitutes one of the most constant differences between the two sexes. These are long, narrow, and pointed in the male; short, broad, and rounded in the female. The tail feathers differ similarly. The sickle feathers and those of the middle row especially continue to grow in the male long after their growth has ceased in the female. Similarly the crest feathers of Polish and Houdans grow longer in the male than in the female, but on account of their greater breadth in the female her crest appears larger and fuller. Lastly, the greater development of the spurs in the male over six to eight months old is a well-marked dimorphic character. Of these characters I have paid most attention to plumage and skin color, and will take up in review the results gained in crossing dimorphic species.

Black Minorca and Dark Brahma.—The male Dark Brahma has white-laced hackles and black, white, and red wing bars. In the female the lacing on the hackles is less conspicuous, and there are no wing bars or bows. Red is wholly absent. All hybrids are prevailingly black. All males, however, show a more or less prominent wing-bar formed of black, straw, and red colored feathers. No females show any trace of a wing bar unless it be a slight iridescence in the wing coverts.

White Leghorn and Dark Brahma.—The male hybrids are typically white, with some red on wing coverts. Apart from some black individuals, the female hybrids are either white, with some buff on wing, or else they resemble the female Dark Brahma, having the penciling modified into mossiness. There is no well-defined wing-bar, but the middle wing is suffused with red.

White Leghorn and Houdan.—Neither of these races exhibits a marked dimorphism in plumage color. Nevertheless, the coloration of the hybrids is dissimilar in the two sexes, the males being of a much purer white than the females (p. 21), and this is true not only in the first generation, but also in the extracted whites of a later generation.

White Leghorn and Rose-comb Black Minorca.—In the first generation the male hybrids were almost without exception pure white; the female hybrids invariably show some black-speckled feathers.

Tosa fowl and White Cochin Bantam.—In the first hybrid generation, as stated, the males had all feathers of male Tosa coloration, but barred with white. The females had the Tosa hen coloration, but with shafting broadened. Here each sex inherits the corresponding characteristic plumage of the Tosa fowl modified by the white of the Cochin, but *in different fashion for each sex*. Barring or cuckoo marking seems, indeed, a prevailingly *male* characteristic. Hurst (1905, p. 133), in crosses of White Leghorn and Houdan, got, in addition to white hybrids, 11 dark birds; of these the 6 pullets were black; the 5 cockerels were barred.

In the F_2 generation I obtained extracted pure (?) male and female Tosa-fowl plumage as well as pure whites (p. 49).

Dark Brahma (female) and Tosa fowl (male).—Here both races are dimorphic. The female hybrids closely resembled in coloration the female Tosa fowl, except that the contour feathers were penciled as in the Brahma. The male hybrids closely resembled in coloration the male Dark Brahma, except that much more red and less white appeared on the wing bars and wing bows. Both sexes inherit some qualities from the corresponding sex of each of the parent species. Again, the males have a yellow foot like their mother, whereas the females have a willow foot like their father. The hybrids of either sex inherit foot-color from the opposite sex of their parents (p. 54).

TRANSFER OF SEXUALLY DIMORPHIC CHARACTERISTICS FROM ONE SEX TO THE OTHER.

Secondary sexual characters, such as have been referred to in the last section, seem indissolubly associated with their corresponding sex. The reason for such an association is obscure, but it is known that it is not due to the absence in the protoplasm of the characteristics of the opposite sex, for these may develop in the individual when the germ glands are removed. The germ glands, then, control the latency of the one set of characters and the patency of the other set. In poultry the removal of the sex glands from a young cock, in the process of caponizing, results in loss of the crowing instinct and failure of comb, wattles, spurs, hackle, saddle, and sickles to acquire the size characteristic of a cock. If in the fowl the germ glands fail to develop, the secondary sex characters are ambiguous.

Despite this apparently physiological dependence of secondary characters in the germ gland, it seems improbable that the association is a necessary one. Almost all characters can be dissociated; why not also sex and secondary characters? There is reason to think much can be done in this way, because something has already been accomplished. For example, the cerebral hernia which now is found equally in both sexes of the Polish fowl was formerly a female secondary sexual characteristic. Bechstein (1793) states that he never observed the cranial dome in male Polish fowl. Blumenbach (1813), who made numerous dissections of the cranium of this fowl, states "of this deformity very slight traces indeed are found in the cocks, and these but seldom."* Consequently it must be concluded that the female secondary characteristic of cerebral hernia has been gradually transferred to the male sex also. A case of which the history is known even more definitely is that of the Sebright Bantam. This bird is characterized by the fact that in the male the hackle, saddle, and sickle feathers are of the same form as in the female; consequently the tail is short and truncate as in a hen. Here, apparently, female characteristics have become attached to a male. Fortunately we have the history of the race from the mouth of the son of the maker, Sir Thomas Sebright. Dr. Horner, who obtained the statement from Sebright, published it in Tegetmeier's Poultry Book (1868, pp. 241, 242).

It was about the year 1800 that the late Sir John Sebright began to fashion the Sebright Bantam. The first cross was between a common Bantam† and the Polish fowl.‡ The chickens resulting from this alliance were bred in-and-in until the required markings and size were secured. Sir John then accidentally found a hen-tailed Bantam cock in the country where he was traveling. This short-tailed bird he in-bred with his newly manufactured Bantams, thereby giving their progeny the present form of the square tail.

The essential characteristic of the race was thus gained from a mutative modification of a polymorphic characteristic.§

In my own experiments I have hardly proceeded far enough to get results; yet already evidence of transference of color characteristics from one sex to the other is appearing. Thus in the second hybrid generation of the Cochin × Tosa cross at least one bird (No. 659 ♀) has hackles of a plain buff color like those of the male Tosa fowl, and entirely unlike the hackles of the female Tosa fowl or the female of the dark variety of the Cochin. Again, the female hybrids between the Dark Brahma hen and White Leghorn cock have much red on the wing coverts. This is foreign to the Dark Brahma hen, and must, so far as I can see, have been derived from the red on the 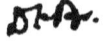 wings of the male ~~Tosa~~ fowl. Finally, two male hybrids between the Tosa fowl and Dark Brahma show the feminine shafting. Experiments in continuance of this investigation are, naturally, in progress.

---

* Translation quoted by Tegetmeier, 1867, p. 173.

† Doubtless Game Bantam is here meant.

‡ The Golden Spangled Polish are undoubtedly referred to, whence the spangling of the feather was obtained. The combination gave the small size and gold-spangled plumage.

§ As might be anticipated from the notoriously sterile quality of hen-feathered cocks, Sebright Bantams are inclined to be sterile, and one is advised not to try to breed from the best show stock, i. e., cocks with the shortest tails (Wright, 1902, p. 598).

SEX IN HYBRIDS.

There is a widely held and frequently expressed opinion that hybrids show an excessive proportion of males. Bateson and Saunders (1902, p. 139) probably have this in mind in their statement—"the statistical distribution of sex among first crosses shows great departure from the normal proportions." I have therefore been interested to tabulate the sex proportions in my hybrids. Without giving the full table, I may state that the totals are: Males, 204; females, 173; sex undetermined, 573. There is here an excess of males; but in view of the large early death rate, this may well be due to a difference in the death rate of the two sexes. Taking the different series of hybrids separately, most of them gave an approximation to equality of the sexes. One of the most striking departures is the series of Dark Brahma (121 ♀) × Tosa (8A ♂) hybrids. Of 22 individuals that developed to 18 days in the incubator, all but one grew to maturity. Of these 21, 16 are males and 5 females. The first egg laid by the Dark Brahma after she was put with the Tosa fowl developed into a female; the next nine that hatched were males; also her last six young were males. The exceptions to the law of equality of sexes in hybrid offspring are thus individual and not of general significance.

CORRELATION OF CHARACTERISTICS.

Every taxonomic description testifies to the fact that a certain set of characteristics is usually found associated in each species or variety. The prevailing theory has been that this association is a necessary one, maintained because all the characters are necessary to the success of the species in its relations to external environment, or else that they were physiologically interdependent. Modern work in hybridizing is establishing the fact that few of the specific characteristics are interdependent. Their association is, so far as interaction goes, mostly accidental. Thus in my experiments with poultry I have merely reached the same conclusions as have been gained by Johannsen (1899, p. 185), de Vries (1903, p. 494), and indeed all recent workers. I find, namely, that of the scores of evident external characteristics of poultry that are inherited in alternative fashion scarcely two can be found that are always associated. The most striking exception is the association of high nostril and absence of single comb.

What, then, is the meaning of correlation in nature? Clearly it is only rarely due to physiological interdependence. It may often be due to an unrelated association of characters independently advantageous to the organism. It is doubtless due to an accidental association of characters brought into the race by successive mutations or by hybridizations and never disturbed, because not prejudicial to the well-being of the species.

THE MUTATION THEORY IN ITS RELATION TO THE ORIGIN OF DOMESTICATED ANIMALS.

While the mutation theory of de Vries has received widespread adherence among botanists, many students of animals, and especially of domesticated races, have appeared as its opponents. Foremost among these are Professors Keller, of Zurich (1905), and Plate (1905), of Berlin. I think that the essence of the mutation theory is too little apprehended. It rests on the fundamental theory of heritable unit characters and assumes their very limited mutability. It recognizes the important results wrought by artificial selection, but considers them as arising from two processes—first, the selection of minute favorable variations of the fluctuating sort, and, secondly, the preservation of new unit characters suddenly appearing. Such unit characters can usually be not only maintained but much improved by subsequent selective breeding.

Now, it is true that breeders nowadays do not regularly wait for favorable qualities to crop out. The process is too slow, uncertain, and expensive. If one had scores of thousands of individuals, desired mutations would come more frequently; but even then they would rarely be of a desirable sort. Every breeder can, on the other hand, improve any characteristic by selection, and that is for the most part the only method of improving a quality that is open to him. Of course he can make new *combinations* of qualities by crossing, but this does not, typically, result in new *qualities*.

The question of the *permanence* of the improvement wrought by selection of minute variations is the first point of difference between de Vries and Keller. De Vries asserts that such improvement persists only so long as selection is maintained. Keller adduces some interesting cases on the other side, and the cogency of some of his evidence must be admitted. He traces the gradual evolution in Egypt of long lop-eared hounds from straight-eared ones. Ear length in rabbits, as Castle (1905, pp. 125-126) has shown, is not a unit character; at least, it blends in hybridization and consequently exhibits any desired intermediate condition. The same is probably due for dogs; consequently this character may well have arisen by summation of minute variations. Yet Keller goes on to show the long-eared condition has persisted in central Africa, where selective breeding no longer occurs. Hence one characteristic originated by selection of fluctuations has not retrogressed on removal of selection.

The preceding method of proof is not, however, of general validity. Evidence that a characteristic arisen in domestication does not disappear when the race becomes feral again is not evidence against the permanence of fluctuations unless it is also proven that the characteristic arose by selection of fluctuations. This is usually not the case. The instance of long ears would seem to be peculiar. Some of the other examples offered by Keller of persistence of characteristics despite discontinuance of selection avail little, since the precise origin of the unit characters concerned is unknown. If, unconsciously or not, a unit character arising as a sport has been preserved under domestication, it will persist even though the race bearing it become feral.

Positive support for the mutation theory is gained from a consideration of the characteristics of poultry. Our study has shown them to be, for the most part, of the order of integral unit characters. As such they could hardly have been "gradually built up." Being indivisible they must have appeared at once, roughly in their present form. The very existence of unit characters is proof of the mutation theory.

That many characteristics of organisms have not been built up, but have suddenly appeared complete, may be inferred from peculiarities of the characters other than their integral nature. For, first, not all kinds of characteristics have been evolved in domestic poultry, but for the most part only such as occur elsewhere among wild races. Thus, for example, booted feet, as found in the grouse; crest on head, as seen in the umbrella bird (*Cephalopterus*), and long tail, as seen also in the widow bird (*Chera*). Secondly, many of the characteristics of domestic poultry are of the order of mutations in so far as they are almost pathological, *e. g.*, taillessness, rose comb, silky and frizzled feathers, cerebral hernia, polydactyl feet, albinism. These characters, cropping out in the sporting organism and not being prejudicial to its well-being, have been preserved by the fancier; they doubtless arose suddenly, as we find arising suddenly to-day other characters, which we discard because incompatible with a healthy stock—such as featherlessness, cross-bill, and imperfect development of toes. If these characteristics appeared suddenly and not by being "built up," as we know is the case, then so, doubtless, have others. The evidence that many, if not most, characteristics of poultry have arisen suddenly, without having been sought and laboriously built up by man, is convincing, and there can hardly be any escape from the conclusion that here evolution has been largely, though not exclusively, by mutation.

E. SUMMARY OF CONCLUSIONS.

(1) Poultry exhibit numerous unit characteristics which do not blend in hybridization, but are inherited in alternative fashion. The unit characters are not immutable things in hybrids, but subject to modification—perhaps permanent—by interaction of the alternative characters.

(2) Although the great majority of characteristics of poultry are inherited alternatively, yet a few cases of color characters show a particulate inheritance. The comparative rarity of blending of characters makes it easier to see how new characters will not be "swamped by intercrossing with the parent form" (page 82).

(3) Specific and varietal characteristics in de Vries's sense are not inherited in a markedly different fashion, although in two cases progressive variants do not Mendelize typically.

(4) The patent characteristic is usually dominant over its latent allelomorph.

(5) Old and new characteristics are equally dominant.

(6) Dominance and recessiveness of characteristics are not always accompaniments of their segregation in the germ cells; both, moreover, are frequently incomplete.

(7) Dominance is usually, but not always, independent of the races crossed.

(8) Prepotency is as truly important in inheritance as dominance.

(9) Many first hybrids exhibit special forms, due to the interaction of the two allelomorphs. These may become fixed as new characteristics.

(10) Reversion is being explained by the persistence in a "latent" condition of the latent character.

(11) An adequate theory of gametic purity has not only to explain the simple Mendelian formula, but also the facts of imperfect dominance, impurity of extracted forms, latency and atavism, and occasional particulate inheritance.

(12) Reciprocal crosses exhibit differences due to the fact that the father and the mother transmit different kinds of characteristics.

(13) When the parent races are dimorphic each sex in the hybrids exhibits the respective sex characteristic of both of the species. In many cases a new form of sexual dimorphism appears in the hybrids.

(14) Certain characteristics of one sex may become transferred to the other by hybridization, owing to lack of complete correlation between primary and secondary sex characters.

(15) The proportion of the two sexes in hybrids is normal.

(16) With few exceptions, correlated characteristics easily separate as a result of hybridization so that any conceivable combination may be effected.

CARNEGIE INSTITUTION,
   STATION FOR EXPERIMENTAL EVOLUTION,
      COLD SPRING HARBOR, *February 12, 1906.*

F. LITERATURE CITED.

AMERICAN POULTRY ASSOCIATION.
    1905. The American standard of perfection. Illustrated. A complete description of all recognized varieties of fowls. Published by Amer. Poultry Assoc. 1905. 299 pp.

BALDAMUS, A. C. E.
    1896. Illustrirtes Handbuch der Federviehzucht. Erster Band: Die Hühnervogel. 3 Aufl. bearbeitet von O. Grünhaldt. Dresden: Schonfeld. 1896. xvi + 476 pp., 102 figs.

BATESON, W., and SAUNDERS, Miss E. R.
    1902. Report I to the Evolution Committee of the Royal Society. London: Harrison. 160 pp.

BATESON, W., and PUNNETT.
    1905. Experimental studies in the physiology of heredity—Poultry. Report II to the Evolution Committee of the Royal Society. pp. 99-131.

BECHSTEIN, J. M.
    1793. Gemeinnütz. Naturgesch. Deutschlands. Bd. 3. Sumpf- u. Hausvögel. Leipzig: Vogel [teste Darwin, 1876].

BLUMENBACH, J. F.
    1805. Handbuch der vergleichenden Anatomie. Göttingen: Dietrich [teste Darwin, 1876].
    1813. De anomalis et vitiosis quibusdam nisis formativi aberrationibus Cum tab. 4°. Göttingen: Dietrich.

BORELLI, P.
    1670. Historiarum et Observationum medicophysicarum Centuriæ IV. Francofurti.

BOVERI, T.
    1902. Ueber mehrpolige Mitosen als Mittel zur Analyse des Zellkerns. Verh. d. phys.-med. Ges. zu Würzburg. N. F. Bd. xxxv.

CANNON, W. A.
    1902. A cytological basis for the Mendelian laws. Bull. Torrey Bot. Club. Vol. 29.

CASTLE, W. E., and ALLEN, G. M.
    1903. The heredity of albinism. Proc. Amer. Acad. of Arts and Sciences. XXXVII, 603-622. April.

CASTLE, W. E.
    1903^c. The heredity of "Angora" coat in mammals. Science, n. s., XVIII, 760, 761. Dec. 11.
    1905. Heredity of coat characters in guinea-pigs and rabbits. Publication No. 23, Carnegie Institution of Washington. Papers of Station for Experimental Evolution No. 2. 78 pp., 6 plates. Feb.

CHAMBERLAIN, B. H.
    1900. Note on a long-tailed breed of fowls in Tosa. Trans. Asiatic Soc. Japan, XXVII.

CLAYTON, J.
    1693. Philosophical Transactions of the Royal Society of London, 1693, p. 992.

CORRENS, C.
    1900. Mendel's Regel über der Verhalten der Nachkommenschaft der Rassenbastarde. Ber. d. deut. bot. Ges , XVIII, 158-168. Sitzung von 27 Apr.
    1905^a. Gregor Mendel's Briefe an Carl Nägeli, 1866-1873. Ein Nachtrag zu den veröffentlichten Bastardierungsversuchen Mendels. Abh. math.-phys. Kl. k. sächs. Ges. d. Wiss., XXIX, No. 3, pp. 189-265.
    1905^b. Einige Bastardierungsversuche mit anomalen Sippen und ihre allgemeinen Ergebnisse. Jahrb. für wiss., Bot. XLI, Hft. 3, pp. 458-484. Taf. V [April].
    1905^c. Zur Kenntnis der scheinbar neuen Merkmale der Bastarde. Bericht der Deutschen Botan. Gesell., XXIII, 70-85.
    1905^d. Über Vererbungsgesetze. Berlin: Borntraeger. 43 pp., 4 figs.

CUÉNOT, L.
　1903. L'hérédité de la pigmentation chez les souris (2ᵐᵉ note). Arch. de zool. expér. et gén. (4) I. Notes et rev., pp. xxxiii–xli.

CUNNINGHAM, J. F.
　1903. Observations and experiments on Japanese long-tailed fowls. Proc. Zool. Soc. London, 1903, I, pp. 227–250.

DARWIN, C.
　1876. The variation of animals and plants under domestication. Second edition, revised. Vols. I, II. New York: D. Appleton & Co.
　[References made, in brackets, to pages of reprint by D. Appleton "Fourth thousand, 1894."]

DAVENPORT, C. B.
　1904. Wonder horses and Mendelism. Science, XIX, 151–152. Jan. 22.

DE VRIES, H.
　1889. Intracellulare Pangenesis. Jena: Fischer. 212 pp.
　1900. Sur la loi de disjonction des hybrides. Compt. Rend. de l'Acad. des Sci. Paris. 26 mars.
　1902. Die Mutationstheorie. Versuche und Beobachtungen über die Entstehung der Arten im Pflanzenreich, Zweiter Band. 1. Lieferung, pp. 1–240.
　1903. Die Mutationstheorie. II. Bd. 2. Lief, pp. 241–496.
　1905. Species and varieties: Their origin by mutation. Ed. by D. T. MacDougal. Chicago: Open Court Publishing Co. 1905. xviii + 847 pp.

DÜRIGEN, B.
　1886. Die Geflügelzucht nach ihrem jetzigen rationellen Standpunkt. Berlin: Parey. 880 pp. 80 Taf u. 101 fig. in text.

FOCKE, W. O.
　1881. Die Pflanzen-Mischlinge. Ein Beitrag zur Biologie der Gewächse. Berlin: Borntraeger. iv + 567.

GALTON, F.
　1883. Inquiries into human faculty. London: Macmillan.
　1889. Natural inheritance. New York and London. ix + 259 pp.

HAACKE, W.
　1893. Gestaltung und Vererbung. Eine Entwickelungsmechanik der Organismen. Leipzig: T. O. Weigel Nachfolger (Tauchnitz). vi + 337.

HÄCKER, V.
　1904. Bastardirung und Geschlechtszellenbildung. Zool. Jahrb., Suppl. VII (Festschrift für Weismann).

HAGENBACH, E.
　1839. Untersuchungen über den Hirn-und Schädelbau der sogenannten Hollenhühner. Archiv. für Anat. Physiol. u. wiss. Med. (Müller) Jg. 1839, pp. 311–331, Taf. XVI.

HURST, C. C.
　1904. Mendel's discoveries in heredity. Trans. Leicester Literary and Philos. Soc., VIII, pp. 121–134. June.
　1905. Experiments with poultry. In Report II to the Evolution Committee of the Royal Society (by Bateson et al.). London: Harrison. 154 pp.

JOHANNSEN, W.
　1899. Sur la variabilité de l'orge considerée au point de vue spécial de la relation du poids des grains à leur teneur en matières azotiques. C. R. trav. de Lab. de Carlsberg. IV. Heft., 4, pp. 122–192.

KELLER, C.
　1905. Die Mutationstheorie von de Vries im Lichte der Haustier-Geschichte. Arch. für Rassen-und Gesellschafts-Biologie. II Jg., 1. Heft., pp. 1–19. Feb.

LANGKAVEL, B.
　1886. Hühner mit sechs Zehen. Der Zoologische Garten., XXVII, p. 35. Jan.

LUCAS, P.
　1847. Traité philosophique et physiologique de l'hérédité naturelle dans états de santé et de maladie du système nerveux., etc. Tom. I. Paris: J. B. Baillière. 1847. 24 + 626 pp.
　1850. [Same title.] Tom. II. Paris: J. B. Baillière. 1850. 936 pp.

McClung, C. E.
    1905. The chromosome complex of orthopteran spermatocytes. Biol. Bull., vol. IX, No. 5, Oct., pp. 304–340.

McGrew, T. F.
    1901. American breeds of fowls. I. The Plymouth Rock. Bull. No. 29, Bureau of Animal Industry, U. S. Dept. Agr. Washington : Government Printing Office.
    1901a. American breeds of fowls. II. The Wyandotte. Bull. No. 31, Bureau of Animal Industry, U. S. Dept. Agr. Washington : Government Printing Office. 30 pp.
    1904. The Shanghai or Cochin fowl. In Weir-Johnson-Brown Poultry Book. pp. 523–558.

Mendel, G.
    1866. Versuche über Pflanzen-Hybriden. Verhandlungen des naturforschen Vereines in Brünn. Bd. IV, 47 pp.

Morgan, T. H.
    1905. The assumed purity of the germ cells in Mendelian results. Science, vol. XXII, No. 574, pp. 877–879. Dec. 29.

Nägeli, C. v.
    1884. Mechanisch-physiologische Theorie der Abstammungslehre.
    1898. A mechanico-physiological theory of organic evolution. Summary. Translation by F. A. Waugh. Chicago : Open Court Publishing Co. 53 pp.

Petersen, C. E.
    1905. The Houdan. In Weir-Johnson-Brown Poultry Book, 1904–05 (q. v.).

Plate, L.
    1905. Die Mutationstheorie im Lichte zoologischer Tatsachen. C. R. du 6me Congrès intern. de Zool. Berne. 1904. 203–212. May 25.

Romanes, G. J.
    1901. Darwin and after Darwin. I. The Darwinian theory (third edition). Chicago : Open Court Publishing Co.

Standfuss, M.
    1896. Handbuch der paläarktischen Gross-Schmetterlinge für Forscher und Sammler. Jena : Fischer. xii + 392. 8 Taf.

Sutton, W. S.
    1902. On the morphology of the chromosome group in *Brachystola magna*. Biol. Bull., IV, No. 1, pp. 24–39. Dec.
    1903. The chromosomes in heredity. Biol. Bull., vol IV, No. 5, pp. 231–251. April.

Tegetmeier, W. B.
    1856. On the remarkable peculiarities existing in the skulls of the feather-crested variety of the domestic fowl, now known as the Polish. Proc. Zool. Soc. Lond., 1856, pp. 366–368. Figures.
    1867. The poultry book, etc. London : Routledge. viii + 356 pp., 30 colored plates, 36 uncolored figures.

Thorndike, E. L.
    1905. Measurements of twins. Archives of philosophy, psychology, and scientific methods. No. 1, Sept. 64 pp.

Tschermak, E.
    1904. Weitere Kreuzungsstudien an Erbsen, Levkojen und Bohnen. Zeitschr. für das landwirths. Versuchswesen in Oesterr. 1904. 106 pp.

Weir, H. ; Johnson, W. G., and Brown, G. O.
    1904–05. The poultry book. 3 vols. New York : Doubleday, Page & Co. xxii + 1311 pp.

Weismann, A.
    1904. The evolution theory. Trans. by J. A. Thomson and M. R. Thomson. 2 vols. London : Arnold. xvi + 416 + 405 pp.

Woods, F. A.
    1902–03. Mental and moral heredity in royalty. Popular Science Monthly. Aug. 1902–April, 1903.

Wright, L.
    1902. The new book of poultry. London, etc. : Cassell & Co. viii + 600 pp.

Wyckoff, E. G.
    1904. The Leghorns. In Weir-Johnson-Brown Poultry Book. 1904–05 (q. v.).

Ziegler, H. E.
    1905. Die Vererbungslehre in der Biologie. Jena : Fischer. 74 pp. 2 Taf.

EXPLANATION OF PLATES.

PLATE I.

FIG. 1.—White Crested Black Polish, ♀ 5. One of the females crossed with the Single-combed Black Minorca (*cf.* fig. 3) to produce the female hybrid shown in fig. 5. (H. A. H.)

FIG. 2.—White Crested Black Polish, ♂ 30. To show the male type of the Polish race, which, when crossed with the Minorca (fig. 4), produces male hybrids like fig. 6. (H. A. H.)

FIG. 3.—Single Comb Black Minorca, ♀ 13. The mother of the hybrids, Minorca × Polish, represented by fig. 5.

FIG. 4.—Single Comb Black Minorca, ♂ 12. The father of various Polish × Minorca crosses, of which a male is represented in fig. 6.

FIG. 5.—First Hybrid between Polish and Minorca, pullet. Compare the females of the parental races, figs. 1 and 3.

FIG. 6.—First Hybrid between Polish and Minorca, cockerel. Compare the males of the parental races, figs. 2 and 4.

The figures marked H. A. H. are from photographs made by Mr. H. A. Hackett.

PLATE II.

FIG. 7.—The head of a Polish fowl, ♂ 3, with skin on left half of head dissected away. Shows cerebral hernia, and the relation to it of the thick skin and crest feathers lying above. Note also the culminal fold, high nostril, and rudimentary comb. The latter lies at the base of the comb and shows as a mottled area against the deep black of the anterior crest feathers. (H. A. H.)

FIG. 8.—Head of a hybrid, ♂ 50, between Minorca ♀ 13 (fig. 3) and Polish ♂ 3 (fig. 7). Shows the Y-shaped comb lying in front of the crest. The comb is double behind, single in front.

FIG. 9.—Head of a Minorca × Polish hybrid of the second generation. The son of such a pair as are represented in figs. 5 and 6. Note the reappearance of a large crest, high nostril, and rudimentary comb. (H. A. H.)

FIG. 10.—Side of cranium of Polish fowl. Shows bony dome covering cerebral hernia.

PLATE III.

FIG. 11.—Head of hybrid of the second generation, Minorca × Polish, 371 ♂—the son of such a pair as are represented in figs. 5 and 6. Note the *absence* (imperfect) of crest, the high nostril, and the rudimentary comb. (H. A. H.)

FIG. 12.—Head of Houdan (♂ 9). Shows crest, high nostril, and rudimentary comb.

FIG. 13.—Foot of Houdan (♂ 9). Shows the two toes (in place of one) situated immediately below the spur.

FIG. 14.—Head of second generation White Leghorn × Houdan hybrid, its father being like fig. 17. Shows the occurrence of both cerebral hernia and single comb on the same individual. (H. A. H.)

PLATE IV.

FIG. 15.—Single Comb White Leghorn (♂ 74). Note high single comb, plain head, clean feet with four toes, and white plumage. (H. A. H.)

FIG. 16.—Houdan (♀ 8). Note crest, high nostril, rudimentary comb, mottled plumage, and muff and beard.

FIG. 17.—First hybrid (♂ 87) between White Leghorn and Houdan. Note crest, Y comb, white plumage, muff, and double toe behind on left foot.

PLATE V.

FIG. 18.—Dark Brahma hen (121). Note uniformity of plumage coloration, except that hackles are *laced* with white, and wing coverts, back, and breast are *penciled*. Comb of pea type. Feet booted. Vulture hock. (H. A. H.)

FIG. 19.—Dark Brahma cock (122). Note laced hackles and saddles, prominent white wing bow, pea comb, and booted feet.

PLATE VI.

FIG. 20.—First generation hybrid, ♂ 607, between Black Minorca (fig. 6) and Dark Brahma (fig. 19). Note prevailingly black plumage, with trace of white wing bow, irregular pea comb, slightly booted feet, and absence of vulture hock and of lacing on hackles. (H. A. H.)

FIG. 21.—First generation hybrid, ♂ 603, between Black Minorca and Dark Brahma. Brother to fig. 20. Note almost complete absence of white wing bow, but presence of white lacing on hackles. Note also high, though pea, comb, and long tail. Form of trunk like Dark Brahma, neck like Minorca. (H. A. H.)

PLATE VII.

FIG. 22.—First generation hybrid, ♀ 387, between White Leghorn Bantam (resembling fig. 15) and Dark Brahma (fig. 18). Note the new type of plumage coloration. Hackles broadly laced as in male, rest of plumage mottled, with much red. Booting rudimentary. (H. A. H.)

FIG. 23.—First generation hybrid, ♀ 395, between White Leghorn and Dark Brahma. Shows reappearance of the Dark Brahma ♀ type of coloration. (H. A. H.)

22

23

PLATE VIII.

FIG. 24.—First generation hybrid, ♂ 270, between White Leghorn and Dark Brahma. Shows the pure white type, i. e., dominance of White Leghorn coloration. Note slight booting, absence of vulture hock and the erect, Leghorn tail. (H. A. H.)

FIG. 25.—First generation hybrid, ♂ 409A, between White Leghorn and Dark Brahma. Shows the type with red on the wing coverts. (H. A. H.)

PLATE IX.

FIG. 26.—Black Cochin Bantam, ♀ 129. Shows short tail and heavily booted feet. The mother of the barred bird, fig. 27. (H. A. H.)

FIG. 27.—First hybrid, ♂ 365, between Black Cochin Bantam (fig. 26) and White Leghorn (*cf.* fig. 15). Note barred plumage coloration, red earlobe, and booted feet. (H. A. H.)

FIG. 28.—Buff Cochin Bantam, ♂ 545. Note short tail, heavily booted feet, red earlobe, and single comb.

PLATE X.

FIG. 29.—Tosa fowl, ♂ 1A, imported from Japan. Long tail feathers had been recently pulled out.

FIG. 30.—Tosa fowl, ♀ 2A, imported from Japan. Note the light shafting.

FIG. 31.—Tosa fowl, ♂ 3A, "Admiral Togo," son of 1A and 2A. Photographed September 7, 1905. Note length of tail. (H. A. H.)

FIG. 32.—White Cochin Bantam, ♀ 35A. This bird was crossed with fig. 29 and gave hybrids represented on plate XI.

PLATE XI.

Fig. 33. First generation hybrid, ♀ 58, between White Cochin (fig. 32) and Tosa fowl (fig. 29). Note a slight broadening of shaft stripe as compared with female Tosa fowl.

Fig. 34.—First hybrid, ♂ 53, between White Cochin and Tosa fowl (fig. 29). Note white barring on feathers, and long tail. (H. A. H.)

Fig. 35.—First hybrid, ♂ 95, between White Cochin and Tosa fowl, younger brother to fig. 34. Note barring and growth of saddle and tail feathers. (H. A. H.)

Fig. 36.—Second generation hybrid, ♀ 312, between White Cochin and Tosa fowl. Note pure white plumage color, like Cochin grandmother, fig. 32, combined with long tail of Tosa, fig. 30. (H. A. H.)

PLATE XII.

FIG. 37.—Plumage chart of F₁ (White Cochin × Tosa), ♀ 58, at about 5 months. pl, to of head; 2, hackle; 3, middle of back; 4, throat; 5, breast; 6, middle tail; 7, saddle; 8, wing, secondary.

FIG. 37a.—Plumage chart of F₁ (White Cochin × Tosa), ♂ 53, at about 5 months. Signification of figures same as in fig. 37. Shows *barring* of feathers.

FIG. 38.—Second hybrid generation (White Cochin × Tosa), ♂ 315. Note reappearance of pure white like Cochin grandmother (fig. 32, plate X); form intermediate, feet booted. (H. A. H.)

37a

PLATE XIII.

FIG. 39.—Jungle fowl, ♀ 2. Taken after death to show shafting on breast, nape, back, and wing coverts.

FIG. 40.—First generation hybrid, ♂ 358, between Dark Brahma (fig. 19, plate V) and Tosa fowl (fig. 29). Note white laced hackles and saddles, vulture hock, boot, and pea comb of Brahma and white earlobe and elongated head of Tosa. (H. A. H.)

39

40

PLATE XIV.

FIG. 41.—Frizzle fowl, ♂ 15. Note rose comb and feathers that turn forward, forming a ruff on the neck. On the exposed vanes of the upper secondaries the twisting of the barbs may be seen.

FIG. 42.—Frizzle fowl, ♀ 18A. Note extreme curling of feathers, the absence of barbs on part of the secondaries, leaving the shaft quite naked, and the absence of plumage on the back of the head.

FIG. 43.—Silky fowl, ♂ 21A. Note single comb, small crest, the downy condition of the contour feathers, and the elongated and disconnected barbs of the wing secondaries and tail feathers.

FIG. 44.—First hybrid between Frizzle and Silky, ♂ 219. Note the white plumage, rose comb, trace of crest, frizzled feathers (ruff !), and absence of elongated barbs on the wing, secondaries, or other evidence of silkiness. The booted feet and extra toe are derived from the Silky. (H. A. H.)

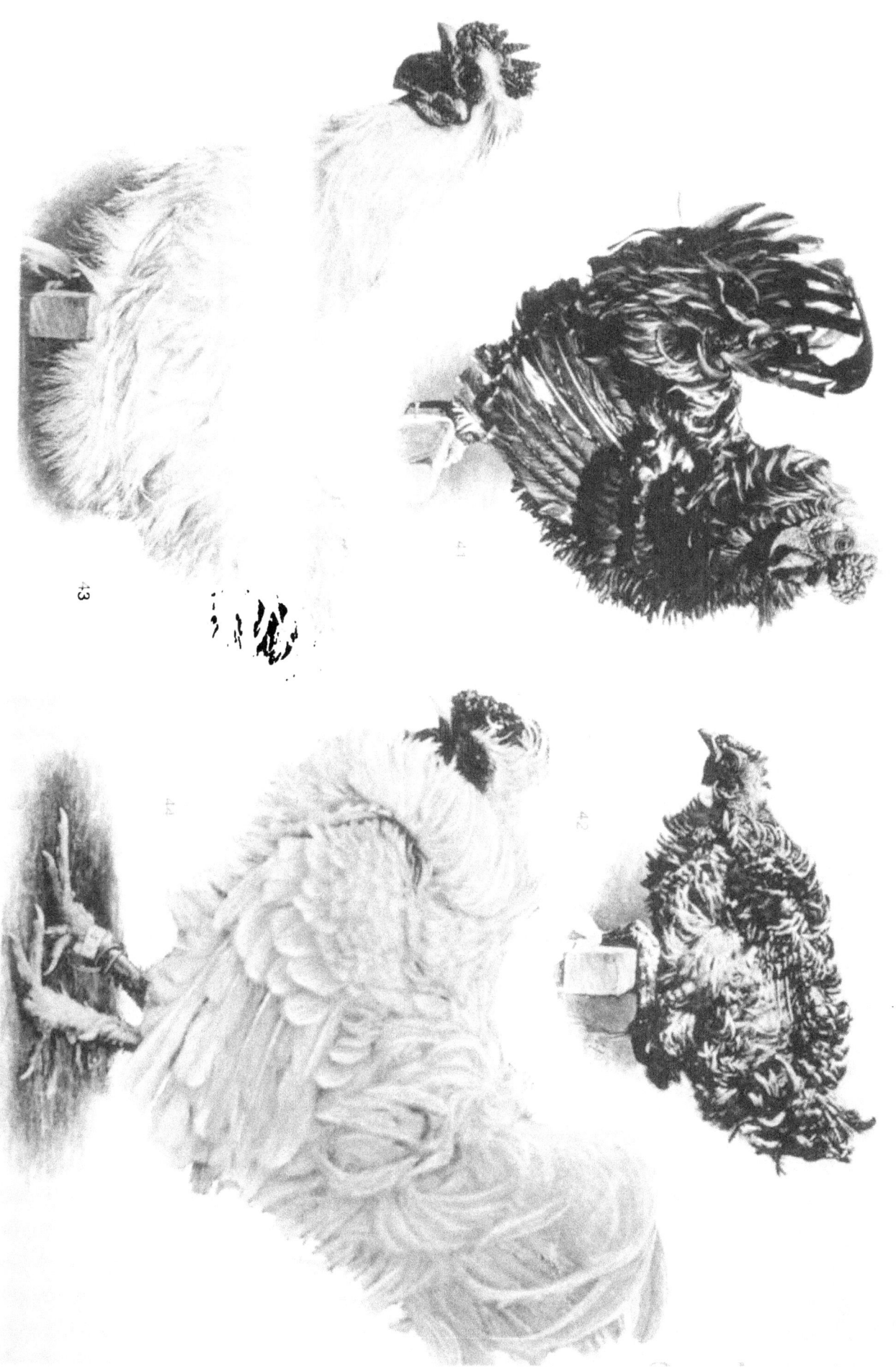

PLATE XV.

FIG. 45.—Rumpless Game, ♀ 49.

FIG. 46.—Rumpless Game, ♂ 117. The hackles and saddles and wing bars are red; otherwise the plumage is largely black. (H. A. H.)

FIG. 47.—First hybrid between White Leghorn (*cf.* fig. 15, plate IV) and Rumpless Game, ♂ 516. Note dominance of white (except for a trace of red on the wing coverts) and long tail. The comb is intermediate between that of a Game and that of a White Leghorn. (H. A. H.)

PLATE XVI.

FIG. 48.—Head of second generation Minorca × Polish hybrid, ♂ 474. Shows last term of series, beginning with fig. 50 and passing through fig. 49, of reduction of median component of Y comb, leaving only two papillae remaining. (H. A. H.)

FIG. 49.—Head of second generation Minorca × Polish hybrid. Shows middle term of series passing from Y comb to V comb. The median portion of the comb is represented by a carunculated mass at the base of the papillae. (H. A. H.)

FIG. 50.—Head of second generation Minorca × Polish hybrid, ♂ 259. Shows beginning degeneration of median component of Y comb, which ends in the V comb (fig. 48). (H. A. H.)

FIG. 51.—Head of second generation Minorca × Polish hybrid. Shows 2 pairs of papillae, high nostrils and rudimentary crest, indicating that the first two characteristics are independant of the third. H. A. H.)

FIG. 52.—Dorsal view of head of (Minorca × Polish) × Minorca hybrid. Shows Y comb in which the median component extends between the arms of the Y, the whole resembling a pea comb. (H. A. H.)

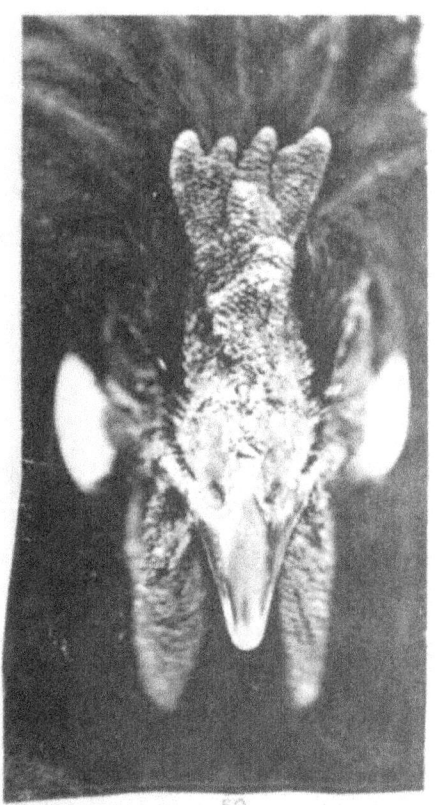
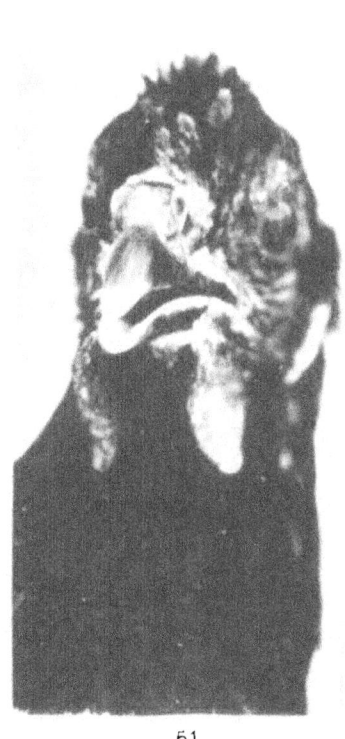

PLATE XVII.

FIG. 53.—First generation hybrid between Silky and Jungle fowl, ♂ 156. Shows dominance of Jungle-fowl plumage color and the extra toe and crest of the Silky. (H. A. H.)

FIG. 54.—First generation hybrid between White Leghorn and Rose Comb Black Minorca, ♀ 138. One of the two birds that exhibit the blue, Andalusian type of coloration, all others being white. (H. A. H.)

53

54

www.ingramcontent.com/pod-product-compliance
Lightning Source LLC
Chambersburg PA
CBHW062323220526
45469CB00008B/2607